Lecture Notes in Mathematics 1511

B. S. Yadav D. Singh (Eds.)

Functional Analysis and Operator Theory

Proceedings of a Conference held in
Memory of U. N. Singh
New Delhi, India, 2-6 August, 1990

Springer-Verlag
Berlin Heidelberg New York
London Paris Tokyo
Hong Kong Barcelona
Budapest

Editors

B. S. Yadav
D. Singh
Department of Mathematics
University of Delhi
Delhi 110007, India

Mathematics Subject Classification (1991): 30D15, 30E10, 41A17, 42A38, 43A25, 43A99, 46A06, 46A35, 46A99, 46B20, 46B99, 46E40, 46J10, 46J20, 46H99, 47A15, 47A17, 47B05, 47B10, 47B37, 47B38, 47D05, 58C20

ISBN 3-540-55365-7 Springer-Verlag Berlin Heidelberg New York
ISBN 0-387-55365-7 Springer-Verlag New York Berlin Heidelberg

Printed in Germany

Typesetting: Camera ready by author
Printing and binding: Druckhaus Beltz, Hemsbach/Bergstr.
46/3140-543210 - Printed on acid-free paper

PREFACE

The department of mathematics of the University of Delhi South Campus, organised a National Seminar from August 2-4, 1990 and an International Conference from August 5-6, 1990 in memory of the late Professor U.N. Singh. The theme of the seminar as well as the conference was `Recent Trends in Contemporary Analysis'. This volume comprises the proceedings of the conference and also includes papers by mathematicians who were unable to attend. The topics that are covered include Functional Analysis, Operator Theory, Abstract Harmonic Analysis, Fourier Analysis Approximation Theory and Function Theory.

The volume is dedicated to the memory of Professor U.N. Singh who was a distinguished analyst and amongst the pioneers who initiated the study of Functional Analysis in India. It includes a paper by him written a few weeks before his demise.

The department is grateful to the University Grants Commission of India and the National Board of Higher Mathematics for their financial support. We would like to express our gratitude to Professor P.K. Jain, Professor R. Vasudevan, Dr. Pramod Kumar, Dr. S.C. Arora and Dr. Ajay Kumar for their help in the preparation of this volume and to Savitri Devi, Poonam Satija and Preeti Nigam for their diligent proof reading.

We are also grateful to Manvinder Singh for placing his computer expertise at our disposal which helped us in various ways.

Finally, we would like to express our thanks to Springer Verlag for their interest and cooperation in bringing out this volume.

B.S. Yadav

Dinesh Singh

CONTENTS

A Qualitative Uncertaininity Principle for Hypergroups

AJAY KUMAR

Dedicated to the memory of U.N. Singh

It is known that for a locally compact abelian group G if $f \in L^1(G)$ and the product of the measure of the support of f and its Fourier transform $\hat{f}$ is less than one then f = 0 a.e. It is also known that if G is with a noncompact identity component and the measure of the support of each f and its Fourier transform $\hat{f}$ is finite, then f = 0 a.e. In this paper we study generalizations of these results for commutative hypergroups.

The uncertainity principles in Fourier analysis assert that the more a function f is concentrated, the more its Fourier transform $\hat{f}$ will be spread out. It has been known for quite sometime that if $f \in L^2(\mathbb{R}^n)$ and the support of f and that of its Fourier transform $\hat{f}$ are bounded, then f = 0 a.e.

For a commutative hypergroup K equipped with Haar measure m, let $\hat{K}$ be the dual space with Plancherel measure π [6] and [7]. For $f \in L^1(K)$, let $A_f=\{x \in K: f(x) \neq 0\}$ and $B_f=\{\gamma \in \hat{K}: \hat{f}(\gamma) \neq 0\}$. The main aim of this paper is to show that if $m(A_f)\pi(B_f) < 1$, then f=0 a.e. If K has a noncompact identity component (with some additional continuity condition) and $m(A_f)<\infty$, $\pi(B_f)<\infty$, then f = 0 a.e.

For $K = \mathbb{R}^n$, the above results have been proved by Matolcsi and Szücs [8] and Benedicks [1] and for locally compact abelian groups by Hogan [5]. Some of the proofs of our results are largely inspired by [5] and [8]. Non abelian groups have been studied by Price and Sitaram [9], Cowling, Price and Sitaram [3] and Echterhoff, Kaniuth and Kumar [4].

Let $B(K)$ denote the σ-algebra of Borel subsets of K and ξ_E the characteristic function of $E \in B(K)$. For $E \in B(K)$ and $F \in B(\hat{K})$, define projections

$P(E)$: $L^2(K) \to L^2(K)$ by $f \to f\xi_E$ and $P^1(F) : L^2(\hat{K}) \to L^2(\hat{K})$ by $\phi \to \phi\xi_F$

Clearly $||P(E)|| \leq 1$ and $||P^1(F)|| \leq 1$. For $f \in L^2(K)$ and $F \in B(\hat{k})$, let Q be the projection on $L^2(K)$ defined by $Q(F)f = (P^1(F)\hat{f})^{\gamma}$. For projections P_1 and P_2 on a Hilbert space H denote by $P_1 \wedge P_2$ the projection on $P_1H \cap P_2H$.

THEOREM 1. If $E \in B(K)$ and $F \in B(\hat{K})$ are such that $m(E)\pi(F) < 1$, then $P(E) \wedge Q(F) = 0$.

PROOF. If $m(E) = 0$. or $\pi(F) = 0$, then for $f \in L^2(K)$, $P(E)f = 0$ or $Q(F)f = 0$, so suppose that $m(E) \neq 0$ and $\pi(F) \neq 0$. Now $m(E)\pi(F) < 1 \Rightarrow m(E) < \infty$ and $\pi(F) < \infty$. Let $g = (\check{\xi}_F)^{\gamma}$. By ([6], 12.1I), it follows that $g \in L^2(K)$. For any $x \in K$,

$$|g(x)| = |\int_K \gamma(x)\xi_F(\gamma)d\pi(\gamma)| \leq \int_F |\gamma(x)| d\pi(\gamma) \leq \pi(F).$$

So for any $x,y \in K$ we have $|g(x*y)| \leq \pi(F) \int_K dp_x * p_y(z) = \pi(F)$.

Now for any $\psi \in L^1(K) \cap L^2(K)$ and $x \in K$,

$$(P(E)Q(F)\psi)(x) = \xi_E(x)\,(P^1(F)\hat{\psi}\,)^{\gamma}(x) = \xi_E(x)(\check{\xi}_F\,\hat{\psi}\,)^{\gamma}(x)$$
$$= \xi_E(x)\,\check{\xi}_F^{\gamma} * \psi(x) = \xi_E(x)\, g*\psi(x)$$

Thus $||P(E)Q(F)\psi||_1 \leq m(E)\pi(F)\, ||\psi||_1$ and

$$||P(E)Q(F)\psi||_2 \leq (m(E))^{1/2}\pi(F)||\psi||_1.$$

Therefore, $P(E)Q(F)\psi \in L^1(K) \cap L^2(K)$. Applying the above inequalities to $P(E)Q(F)\psi$, we get

$$||(P(E)Q(F))^2\psi||_1 \leq (m(E))^2(\pi(F))^2 \quad ||\psi||_1$$

and $\quad ||(P(E)Q(F))^2\psi||_2 \quad \leq (m(E))^{3/2}(\pi(F))^2||\psi||_1.$

Thus by inducton

$$||(P(E)Q(F))^n\psi||_2 \leq (m(E))^{n-1/2}(\pi(F))^n ||\psi||_1.$$

Therefore $||(P(E)Q(F))^n\psi||_2 \to 0$ as $n \to \infty$.

Let $f \in L^2(K)$.There exists $f_1 \in L^1(K) \cap L^2(K)$ such that $||f-f_1|| < \epsilon/2$.

Choose n large enough so that $||(P(E)Q(F))^n f_1|| < \epsilon/2$. Hence

$$||(P(E)Q(F))^n f||_2 \leq ||(P(E)Q(F))^n(f-f_1)||_2 + ||(P(E)Q(F))^n f_1||_2 < \epsilon.$$

So $||(P(E)Q(F))^n f||_2 \to 0$ as $n \to \infty$ for all $f \in L^2(K)$.

Thus for $f \in P(E)(L^2(K)) \cap Q(F)(L^2(K))$, we have

$$||f||_2 = ||(P(E)Q(F))^n f||_2 \to 0 \text{ so } f = 0.$$

COROLLARY 2. Let $f \in L^1(K)$ be such that $m(A_f)\pi(B_f) < 1$.Then $f = 0$ a.e.

PROOF. We first remark that $f \in L^1(K)$, $m(A_f) < \infty$ and $\pi(B_f) < \infty$ if and only if $f \in L^2(K)$, $m(A_f)<\infty$ and $\pi(B_f)<\infty$.Clearly $P(A_f)f= f$ and $Q(B_f)f = (P^1(B_f)\hat{f})^{\vee} = (\hat{f})^{\vee} = f$ by ([6], 12.2C). So $P(A_f)Q(B_f)f= f$. Hence $||f|| = ||(P(A_f)Q(B_f))^n f|| \to 0$, and hence $f = 0$ a.e.

Now we proceed to prove the following version of the qualitative uncertainity principle (QUP). K is said to satisfy QUP if for each $f \in L^1(K)$, $m(A_f)<\infty$ and $\pi(B_f)<\infty \Rightarrow f = 0$ a.e.

LEMMA 3. If C is a compact subset of K, then the map $a \to m(a * C)$ is continuous.

PROOF. For $a \in K$, $a*C$ is compact. By regularity of the Haar measure, we have for $\epsilon>0$, there exists an open set W such that $a*C \subset W$ and $m(W)<m(a*C)+\epsilon$. Now for every $c \in C_1 = a*C$ there exists a neighbourhood U_c of e such that $c * U_c \subset W$. Also there exists a neighbourhood V_c of e such that $V_c*V_c \subset U_c$. Now $\{ c*V_c : c \in C_1\}$ covers C_1, therefore, there exists a finite subcover

$\{ c_i*V_{c_i} : 1 \leq i \leq n\}$ of C_1. Let $V = \bigcap_{i=1}^{n} V_{c_i}$. Then

$$a*C*V \subset \bigcup_{i=1}^{n} c_i * V_{c_i} * V \subset \bigcup_{i=1}^{n} c_i * V_{c_i} * V_{c_i} \subset \bigcup_{i=1}^{n} c_i * U_{c_i} \subset W.$$

Thus $m(a*C*V) \leq m(W) < m(a*C)+\epsilon$. Hence by using ([6],3.3C) it follows that the map $a \to m(a*C)$ is continuous.

Using the above lemma, it follows easily that if K is a compact hypergroup, a discrete hypergroup or a locally compact group and $C \subseteq K$ is such that $0 < m(C) < \infty$, then the map $a \to m(a*C)$ is continuous. We don´t know whether the above lemma can be extended to a subset C of K with $0<m(C)<\infty$. However, we assume in the remaining part that K satisfies the condition that $a \to m(a*C)$ is continuous for every $C \subseteq K$ with $0<m(C)<\infty$.

PROPOSITION 4. Let K be a commutative hypergroup with anoncompact identity component K_o. Let C be a measurable subset of K with $0<m(C)<\infty$. If $C_o \subseteq C (m(C_o)>0)$ and $\epsilon>0$, then there exists $a \in K_o$ such that

$$m(C) < \int_K (\xi_C + \xi_{a*C_o} - \xi_C \, {}_{\tilde{a}}(\xi_{C_o}))dm < m(C) + \epsilon.$$

PROOF. Define $h : K_o \to \mathbb{R}^+$ by $h(a) = \int_K (\xi_C + \xi_{a*C_o} - \xi_C \,{}_{\tilde a}(\xi_{C_o}))\, dm$ so that $h(a) = m(C) + m(a*C_o) - \langle {}_a(\xi_{C_o}), \xi_C \rangle$. By assumption and the continuity of the map $a \to {}_a f$ ([2], §2), it follows that h is a continuous function. Select $\delta > 0$ such that $0 < 2\delta < m(C_o)$. There exists a compact set $F \subseteq C$ such that $m(C \sim F) < \delta$. Let $M = F * \tilde{F}$, by ([6],3.2B) M is a compact subset of K. Since $e \in M$ and $e \in K_o$, $M \cap K_o$ is compact. As K_o is non-compact, select $a \in K_o \sim (M \cap K_o)$. It is easy to see that ${}_a(\xi_F)\xi_F = 0$. If $x \notin F$, then clearly ${}_a(\xi_F)\xi_F(x)=0$. If $x \in F$, then

${}_a(\xi_F)\xi_F(x)=\xi_F(a*x) = 0$, since $\{a\}*\{x\}\cap F = \phi$. In fact, if $y \in \{a\}*\{x\}\cap F$, then $a \in \{\tilde{x}\}*\{y\} \subseteq \tilde{F} * F = M$, and hence $a \in M \cap K_o$, which is a contradiction. Next we claim that

$\int_K {}_{\tilde a}(\xi_{C_o})\, \xi_F\, (x)dm(x) < \delta$. In fact,

$$\int_K {}_{\tilde a}(\xi_{C_o})\, \xi_F(x)\, dm(x) =$$

$$\int_K {}_{\tilde a}(\xi_{C_o\cap F} + \xi_{C_o\cap F'})(x)\, \xi_F(x)dm(x)$$

$$= \int_K \xi_{C_o}(x)\, \xi_F(x)\, {}_a(\xi_F)(x)dm(x) + \int_K \xi_{C_o\cap F'}\, {}_a(\xi_F)dm \quad ([6],\ 5.1D)$$

$$= \int_K \xi_{C_o\cap F'}\, {}_a(\xi_F)dm \le \int_K \xi_{C_o\cap F'}\, dm = m(C_o\cap F')$$

$$\le m(C \sim F) < \delta.$$

Thus

$$h(a) = \int_K (\xi_C + \xi_{a*C_o} - \xi_C \tilde{_a}(\xi_{C_o}))(x)dm(x)$$

$$= m(C) + m(a*C_o) - \int_K {_a}(\xi_{C\cap F} + \xi_{C\cap F'})\, \xi_{C_o}\, dm \qquad ([6],5.1\ D)$$

$$\geq m(C) + m(C_o) - \int_K {_a}(\xi_F)\, \xi_{C_o}\, dm - \int_K \xi_{C\cap F'}\, \tilde{_a}(\xi_{C_o})dm \quad (\xi_{a*C_o} \geq \tilde{_a}(\xi_{C_o}))$$

$$\geq m(C) + m(C_o) - \int \xi_F\, \tilde{_a}(\xi_{C_o})dm - \int \xi_{C\cap F'}\, dm$$

$$> m(C) + m(C_o) - (m(C_o)/2) - (m(C_o)/2 = m(C) = h(e).$$

Hence h is a nonconstant continuous function on the connected set K_o and $h(a)>h(e)$. We may now choose $a_o \in K_o$ such that

$$m(C)=h(e) < h(a_o) = \int_K (\xi_C + \xi_{a*C_o} - \xi_C \tilde{_{a_o}}(\xi_{C_o}))(x)dm(x)$$

$$< h(e) + \in = m(C) + \in.$$

THEOREM 5. If K is a commutative hypergroup with a noncompact identity component such that the map $a \to m(a*C)$ is continuous for all $C \subseteq K$ with $o < m(C) < \infty$, then K satisfies QUP.

PROOF. Let $f \in L^1(K)$ be such that $m(A_f) < \infty$ and $\pi(B_f) < \infty$. Suppose $f_o \in P(A_f)(L^2(K))\cap Q(B_f)(L^2(K))$ and $f_o \neq 0$. Let $A_o = \{x \in K : |f_o(x)| > 0\}$ so $m(A_o) > 0$. Select $N \in \mathbb{N}$ with $2m(A_o)\pi(B_f) < N$. Take $C_o = A_o$ and $C = A_o$ in the above proposition.

There exists $a_1 \in K_o$ such that

$$m(A_o) < \int_K (\xi_{A_o} + \xi_{a_1*A_o} - \xi_{A_o} \tilde{_{a_1}}(\xi_{A_o}))(x)dm(x) < m(A_o) + 1/2\pi(B_f).$$

Take $C_o = A_o$ and $C = A_o \cup (a_1 * A_o) = A_1$ in the above proposition. Then there exists an $a_2 \in K_o$ such that

$$m(A_1) < \int_K (\xi_{A_1} + \xi_{a_2 * A_o} - \xi_{A_1}\, \tilde{a}_2(\xi_{A_o}))dm < m(A_1) + 1/(2\pi(B_f))$$

Repeating the above process we get $A_i = A_{i-1} \cup (a_i * A_o)$ with $a_i \in K_o$ satisfying

$$m(A_{i-1}) < \int_K (\xi_{A_{i-1}} + \xi_{a_i * A_o} - \xi_{A_{i-1}}\, a_i(\xi_{A_o}))dm < m(A_{i-1}) + 1/(2\pi(B_f))$$

As in the proof of Theorem 1, $P(A_i)Q(B_f)\varphi(x) = \xi_{A_i}(x)(\xi_{B_f})^{\gamma} * \varphi(x)$.

Using ([5],1.2), it follows that

$$||P(A_i)Q(B_f)||_2^2 \le \int_K \int_K |\xi_{A_i}(x)(\xi_{B_f})^{\gamma}(x * \tilde{y})|^2\, dm(y)\, dm(x)$$

$$= \int_{A_i} \int_K |{}_x((\xi_{B_f})^{\gamma})(\tilde{y})|^2\, dm(y)dm(x)$$

$$= \int_{A_i} \int_{\hat{K}} |\gamma(x)\, \xi_{B_f}(\gamma)|^2\, d\pi(\gamma)dm(x)$$

$$\le m(A_i)\pi(B_f).$$

Thus $\dim (P(A_N)(L^2(K)) \cap Q(B_f)(L^2(K)) \le m(A_N)\pi(B_f)$

$$= m(A_{N-1} \cup (a_N * A_o)\pi(B_f)$$

$$= \int_K (\xi_{A_{N-1}} + \xi_{a_N * A_o} - \xi_{A_{N-1}}\, \xi_{a_N * A_o})dm\, \pi(B_f)$$

$$\le \int_K (\xi_{A_{N-1}} + \xi_{a_N * A_o} - \xi_{A_{N-1}}\, \tilde{a}_N(\xi_{A_o}))dm\, \pi(B_f)$$

$< (m(A_{N-1}) + 1/(2\pi(B_f))\pi(B_f)$

$\vdots$

$< (m(A_o) + N/(2\pi(B_f))\pi(B_f) < N/2 + N/2 = N.$ ———————— (1)

Let $f_i = {}_{a_i}(f_o)$ so that $\hat{f}_i(\gamma) = \gamma(a_i)\hat{f}_o(\gamma)$.

Now $Q(B_f)f_i(x) = \int_{\hat{K}} \xi_{B_f}(\gamma)\gamma(a_i * x)\,\hat{f}_o(\gamma)d\pi(\gamma)$

$$= Q(B_f)f_o\,(a_i * x) = f_o(a_i * x) = f_i(x)$$

for all $0 \leq i \leq N$.

As $\operatorname{supp} f_i = \{a_i\} * \operatorname{supp} f_o$, $f_i = 0$ a.e. on $(a_i * A_o)$.

Since $P(A_m)f_i(x) = \xi_{A_m}(x)\,f_o(a_i * x)$, we have

$P(A_m \sim A_{m-1})\,f_m(x) = \xi_{A_m \sim A_{m-1}}(x)\,f_o(a_m * x) \neq 0$

and $P(A_m \sim A_{m-1})\,f_i(x) = 0$ for $0 \leq i \leq m-1$. Therefore f_m is not a linear combination of $f_o, f_1 \ldots f_{m-1}$ and hence $\{f_o, f_1 \ldots f_N\}$ is a set of N+1 linearly independent set in $P(A_N)(L^2(K)) \cap Q(B_f)(L^2(K))$. This leads to a contradiction to (1). Hence $f_o = 0$ a.e.

I would like to thank Prof.(Mrs.) Ajit I. Singh for some useful discussions.

REFERENCES

1. M. Benedicks, Fourier transforms of functions supported on sets of finite Lebesgue measure, J. Math. Anal. Appl. 106(1985), 180-183.

2. A.K. Chilana and K.A. Ross, Spectral synthesis in hypergroups, Pacific J. Math. 76(1978), 313-328.

3. M.Cowling, J.F. Price and A.Sitaram, A qualitative uncertainity principle for semisimple Lie groups, J. Austral. Math. Soc. 45(1988), 127-132.

4. S. Echterhoff, E. Kaniuth and A. Kumar, Qualitative uncertainity principle for certain locally compact groups, Forum Mathematicum 3(1991),355-369.

5. J.A. Hogan, A qualitative uncertainity principle for locally compact abelian groups, Proc. Centre Math. Anal. Austral Nat. Univ. 16(1988), 133-142.

6. R.I. Jewett, Spaces with an abstract convolution of measures, Advances in Math. 18(1975), 1-101.

7. G.L. Litvinov, Hypergroups and hypergroup algebra J.Sov. Math. 38(1987), 1734-1761.

8. T. Matolcsi and J. SzÜcs, Intersection des measures spectrales conjugees C.R. Acad. Sci. Paris Ser. A277 (1973), 841-843.

9. J.F.Price and A.Sitaram, Functions and their Fourier transforms with supports of finite measure for certain locally compact groups, J. Funct. Anal. 79(1988) 166-182.

Department of Mathematics,
Rajdhani College,
(University of Delhi),
Raja Garden,
Delhi - 110015.

Weighted Shifts and Composition Operators on L^2

ALAN LAMBERT

THIS ARTICLE IS DEDICATED TO THE MEMORY OF PROFESSOR U.N.SINGH

Let l_+^2 be the Hilbert space of complex valued, square summable sequences, and let $\{e_n; n\geq 0\}$ be the standard orthonormal basis for this space. A wieghted shift is an operator S on l_+^2 defined by the system of equations $[Se_n=\alpha_{n+1}e_{n+1}]$, where $\{\alpha_n\}$ is a bounded sequence of positive numbers. Shifts are discussed in depth in Allen Shield's survey article [7]. In the present article we shall compare certain properties of shifts with those of composition operators. Given a σ-finite measure space (X,Σ,m) and a measurable transformation $T:X \to X$ such that $m \circ T^{-1} << m$, we define C to be the composition operator induced by T acting on $L^2(X,\Sigma,m)$ by $Cf=f\circ T$. C acts as a bounded operator so long as $\frac{dm \circ T^{-1}}{dm} \in L^\infty$. This boundedness condition will be assumed to hold through out this article. In order to see the relation between shifts and composition operators we shall view l_+^2 as $L^2(\mathbb{Z}^+)$. Moreover we shall employ the atomic measure on $\mathbb{Z}^+$ whose mass at the nonnegative integer k is

$$m(k)=\left(\prod_{j=1}^{k}\alpha_j\right)^{-2},\ m(0)=1.$$

It follows that for the shift S mentioned above, S^* is unitarily equivalent to the composition operator on $L^2(\mathbb{Z}^+,m)$ induced by the transformation $T:0\to 1\to 2\to \ldots.$

Note that from a measure-theoretic point of view this type of transformation is perhaps the simplest example of a non-invertible transformation. We shall see that it is this simplicity that distinguishes shift adjoints among the class of composition operators. It will prove helpful to gather some of the basic terminology and properties of composition operators.

. $h_n = \dfrac{dm\ 0\ T^{-n}}{dm}$

. E_n is the conditional expectation operator with respect to $T^{-n}\Sigma$:

$E_n(f)=E(f|T^{-n}\Sigma)$.

. All set and function statements are to be interpreted as holding up to sets of measure 0.

. H_n is the support of h_n.

. $\Sigma_\infty = \bigcap_{n=1}^{\infty} T^{-n}\Sigma$

We shall repeatedly make use of the following general properties of measurable transformation.

. Each $T^{-1}\Sigma$-measurable function F has the form foT for some measurable function f. If foT = goT then f=g on H. In particular, even though T may fail to be invertible, the notation $h.(Ef)oT^{-1}$ is justified. In fact, $C^*f = h.(Ef)oT^{-1}$. (See [1] and [5])

. $\{H_n\}$ is a decreasing sequence of sets. ([3])

. $T^{-1}H=X$. ([3])

. $h_{n+1}=h.(Eh_n)oT^{-1}=h_n.(E_nh)oT^{-n}$. ([5])

Our first illustration of an operator-theoretic property which is essentially automatic for shifts but is not so for composition operators is that of being centered. In 1974 B. Morrell and P. Muhly ([6]) introduced this concept. A is centered if $\{A^{*n}A^n, A^mA^{*m}: n,m\geq 0\}$ is commutative. This property is enjoyed by all normal operators. For a weighted shift, all of the operators involved in the definition of centredness are given by diagonal matrices, and so every weighted shift is centered. Now an operator A is centered if and only if A^* is centered. It might seem reasonable to conjecture that every composition operator is centered. However this is not the case.

1. PROPOSITION. ([3]) The composition operator C induced by T is centered if and only if h is Σ_∞-measurable.

To see how the weighted shift case follows from Proposition 1, note that for each nonnegative integer k, $\{k\}=T^{-1}(\{k+1\})$, so that $T^{-1}\Sigma=\Sigma$. From this it readily follows that $\Sigma_\infty=\Sigma$, and consequently any choice of weights leads to a centered operator. On the other hand, to see that there are composition operators which are not centered, consider the following ``shift-like'' composition operator:

Let $X=\mathbb{Z}^+$. The measure m is given by the counting measure

$m(k)=1$, $k=0,1,2....$

Finally, the transformation T is given by:

$T(0)=0$; $T(k+1)=k$, $k\geq 1$.

Now, one readily sees that for each positive integer n, $T^{-n}\{0\}=\{0,1,...n\}$ and $T^{-n}\{k\}=\{k+n\}$ for $k>0$. It then follows that $\Sigma_\infty=\{\phi,X\}$ and so the only Σ_∞-measurable functions are the constant functions. However, for the transformation T given above, $h(0)=\dfrac{m(0)+m(1)}{m(0)}=2$ and $h(1)=\dfrac{m(2)}{m(1)}=1$. In particular, h is not Σ_∞-measurable and consequently C is not centered.

Remark. The previous example indicated a potentially interesting area of investigation, namely essentially centered operators; that is, operators A for which $\{\pi(A^{*n}A^n),\pi(A^mA^{*m}): n,m\geq 0\}$ is an abelian subset of the Calkin algebra (Here we are taking π to be the canonical homomorphism onto the Calkin algebra.). The composition operator given is relevant since it is a rank one perturbation of a weighted shift.

We are about to examine composition operator adjoints and shifts with relation to the property of hyponormality. Recall that an operator A is hyponormal if $A^*A \geq AA^*$. It is well known that the weighted shift described at the beginning of this article is hyponormal if and only if $\{\alpha_n\}$ is monotonically increasing. Now, there are many examples of hyponormal operators with the property that some power of the operator fails to be hyponormal. Indeed, there are even examples of composition operators with this property ([2]);([5]). However one need only investigate shifts on an elementary level to see that if S is a hyponormal weighted shift, then S^n is hyponormal for each $n \geq 1$. In this regard the adjoint of a composition operator acts very much like a shift, i.e., if C^* is hyponormal, then C^{*n} is hyponormal for each $n \geq 0$. The key to this fact is the elegant characterization of hyponormality for C^* developed by D. Harrington and R. Whitley:

2. PROPOSITION. **C^* is hyponormal if and only if every measurable subset of H is in $T^{-1}\Sigma$, and $h \circ T \geq h$ a.e.**

(It is an interesting exercise to see how this Proposition applies to weighted shifts.)

Using Proposition 2, it is shown in [3] that if C^* is hyponormal then $\Sigma_\infty = T^{-1}\Sigma$ and C is centered. From this it readily follows that if C^* is hyponormal, then all so are of its powers.

Unlike centeredness and hyponormality, the third and last basic property we shall investigate with regard to shifts and composition operators is not an operator-theoretic concept. It is an intgral part of the analysis of measurable transformations: ergodicity. A transformation T is said to be ergodic if for every measurable set A with $T^{-1}A=A$, $m(A)=0$ or $m(X-A)=0$. In the case of a weighted shift the transformation in indeed ergodic. For arbitrary composition operators this need not be the case. However when T is ergodic and C^* is hyponormal, C^* is in fact a weighted shift in a somewhat more general sense than defined previously. To be precise, let $\{H_n\}_{n=0}^{\infty}$ be a sequence of Hilbert spaces and let $H=\Sigma\oplus H_n$. Let $\{A_n\}$ be a sequence of uniformily bounded linear transformations, where each A_{n+1}

$: H_n \to H_{n+1}$. Then the operator S on H given by $S\langle x_0, x_1, \ldots\rangle = \langle 0, A_1 x_0, A_2 x_1, \ldots\rangle$ is called an operator weighted shift.

We note that for any weighted shift, when viewed as the adjoint of a composition operator we have m(X-H)>0. (Indeed, X-H={0}.)

3. PROPOSITION. Let C be the composition operator induced by T and suppose that m(X-H)>0, C^* is hyponormal, and T is ergodic. Then C^* is unitarily equivalent to the direct sum of the zero operator and an operator weighted shift.

Proof. Since C^* is hyponormal, we have $\Sigma_\infty = T^{-1}\Sigma$. Moreover, the kernel of C^* is a reducing subspace for C^*. But this subspace is precisely the orthogonal complement of $L^2(T^{-1}\Sigma)$. Thus by restricting C to the closure of its range (which we have just indicated is a reducing subspace) we may assume that $T^{-1}\Sigma=\Sigma$. In particular we may write X-H in the form $T^{-1}K_1$. Beside $T^{-1}H=X$ we may choose $K_1 \subset H$. Continuing this process, we arrive at a chain

$$K_o = X-H = T^{-1}K_1 = T^{-2}K_2 = \ldots$$

where for each n, $K_n \subset H_n$ and $T^{-1}K_{n+1}=K_n$. Now, $K_o \neq \phi$, $T^{-1}K_o=\phi$, and so $T^{-1}\bigcup_{n=0}^{\infty} K_n = \bigcup_{n=0}^{\infty} K_n$. It then follows from the assumption of ergodicity that $\bigcup_{n=0}^{\infty} K_n$ = X. However a routine induction argument shows that the K_n's are pairwise disjoint. Thus $L^2(X)=\Sigma\oplus L^2(K_n)$. It is simple matter to verify that $C^*L^2(K_o)$

$= 0$ and $C^*L^2(K_{n+1}) \subset L^2(K_n)$, completing the proof.

Although this note has dealt almost exclusively with C^*, there is a substantial number of publications relating to normality, quasinormality, and subnormality for C which have not been referenced explicitly in the article. Some of these are listed here, preceding the references for this article.

Related Readings:

M. Embry-Wardrop, Subnormal centered composition operators, preprint.

T. Hoover and A. Lambert, Essentially normal composition operators, preprint.

A. Lambert, Subnormality and weighted shifts, J. London Math. Soc. (2) 14 (1976), 476-480.

A. Lambert, Subnormal composition operators, Proc. Amer. Math. Soc., 103, no. 3 (1988), 750-754.

A. Lambert, Normal extensions of subnormal composition operators, Michigan Math. Jour., 35 (1988), 443-450.

A. Lambert, A bi-measurable bijection generated by a non-measure preserving transformation, to appear in Rocky Mt. J. Math.

E. Nordgren, Composition operators in Hilbert space, Hilbert space operators, Lecture Notes in Math., vol. 639, Springer-Verlag, Berlin and New York, 1978.

R. Singh, A. Kumar, and D. Gupta, Quasinormal composition operators on l^2_p, Indian Jour. of Pure and Appl. math., 11 (7) (1980), 904-907.

R. Singh and A. Kumar, Characterizations of invertible unitary, and normal composition operators, Bull, Austral. Math. Soc., 19(1978), 81-93.

R. Singh and T. Veluchamy, Non atomic measure spaces and Fredholm composition operators, Acta Scient. Math. (Szeged), 51(1987), 461-465.

R. Whitley, Normal and quasinormal composition operators, Proc. Amer. Math. Soc., 70(1978), 114-118.

REFERENCES

1. J.Campbell and J. Jamison, On some classes of weighted composition operators, to appear in Glasgow Math. Jour.

2. P. Dibrell and J. Campbell, Hyponormal powers of composition operators, Proc. Amer. math. Soc., 102(1988), 914-918.

3. M.Embry-Wardrop and A. Lambert, Measurable transformations and centered composition operators, preprint.

4. D. Harrington and R. Whitley, Seminormal composition operators, J. Oper. Theory, 11 (1984), 125-135.

5. A. Lambert, Hyponormal composition operators, Bulll. London Math. Soc. 18(1986), 395-400.

6. B. Morrell and P. Muhly, Centered operators, Studia Math. 51(1974), 251-263.

7. A. Shields, Weighted shift operators and analytic function theory, Topics in Operator Theory, Math. Surveys, no. 13, Amer. Math. Soc., Providence, R. I., 1974.

Department of Mathematics
University of North Carolina at Charlotte, Charlotte
N.C. 28223, U.S.A.

VARIATION DIMINISHING PROPERTIES AND CONVEXITY FOR THE TENSOR PRODUCT BERNSTEIN OPERATOR

A.S. CAVARETTA, JR. AND A. SHARMA

DEDICATED TO THE MEMORY OF PROFESSOR U.N. SINGH

1. Recently Chang and Davis [1] have given a necessary and sufficient condition for the convexity of Bernstein polynomials over the triangle. Later Goodman [5] studied the variation diminishing properties of Bernstein polynomials over the triangle. Here we consider corresponding results for tensor product Bernstein polynomials.

For positive integers n,m and for $0 \le x \le 1$, we set

$$B_{n,m}f(x,y) = \sum_{i=0}^{n} \sum_{j=0}^{m} f_{ij} x^i y^j (1-x)^{n-i} (1-y)^{m-j} \binom{n}{i}\binom{m}{j} \tag{1.1}$$

where $f_{ij} := f\left(\frac{i}{n}, \frac{j}{m}\right)$, $i = 0,1,\ldots,n$; $j=0,1,\ldots,m$.

We now introduce the function $\hat{f} := \hat{f}_{n,m}$ called the ``net'' determined by the values f_{ij}. A grid determined by the lines $x = i/n$, $i=0,1,\ldots,n$ and $y = j/m$ $j=0,1,\ldots,m$ is placed on the unit square yeilding mn rectangles R_{ij} with vertices

$$\left(\frac{i}{n}, \frac{j}{m}\right), \left(\frac{i+1}{n}, \frac{j}{m}\right), \left(\frac{i+1}{n}, \frac{j+1}{m}\right) \text{ and } \left(\frac{i}{n}, \frac{j+1}{m}\right).$$

On R_{ij} we define the restrictions of $\hat{f} = \hat{f}_{nm}$ by

$$\hat{f}|_{R_{ij}} := f_{ij}(i+1-nx)(j+1-my) + f_{i+1,j}(nx-i)(j+1-my) + f_{i,j+1}(my-j)(i+1-nx) + f_{i+1,j+1}(nx-i)(my-j). \tag{1.2}$$

Clearly $\hat{f}$ is continuous, interpolates f at the grid points, and is bilinear on each rectangle R_{ij}.

Following Goodman, we define the variation of f when $f \in C^2(R)$ by

$$V(f,R) = \iint_R (f_{xx}^2 + 2f_{xy}^2 + f_{yy}^2)^{1/2}\, dxdy \tag{1.3}$$

where R is any rectangle with sides parallel to the axes.

This definition does not make sense for $\hat{f}$ since $\hat{f}$ has discontinuous first derivative. The variation of $\hat{f}$ consists of two parts. In the interior of each R_{ij}, we have a contribution to $V(\hat{f},R)$ given by

$$V^o(\hat{f};R_{ij}) := \sqrt{2}\iint_{R_{ij}} |\hat{f}_{xy}|\,dxdy.$$

Across the line segment joining $\left(\frac{i}{n},\frac{j}{m}\right)$ and $\left(\frac{i}{n},\frac{j+1}{m}\right)$ seperating $R_{i-1,j}$ and R_{ij}, we compute the change in the gradient of $\hat{f}$.

If we set

$$\delta_1 f_{ij} := f_{i+1,j} - 2f_{ij} + f_{i-1,j}$$

(1.3a)

$$\delta_1 f_{ij} := f_{i,j+1} - 2f_{ij} + f_{i,j-1}$$

then the change in the gradient of $\hat{f}$ across the line segment $\left(\frac{i}{n},y\right)$, $\frac{j}{m} \le y \le \frac{j+1}{m}$, is

$$n\{(j+1-my)\delta_1 f_{ij} + (my-j)\delta_1 f_{i,j+1}\},\quad \frac{j}{m} \le y \le \frac{j+1}{m}.$$

Hence we define

(1.4) $$V^1_{ij}(\hat{f}) := n\int_{j/m}^{j+1/m} |(j+1-my)\delta_1 f_{ij}+(my-j)\delta_1 f_{i,j+1}|\,dy$$

with an analogous formula for $V^2_{ij}(\hat{f})$. Putting $my-j = u$, we see that

$$V^1_{ij}(\hat{f}) = \frac{n}{m}\int_0^1 |(1-u)\delta_1 f_{ij} + u\delta_1 f_{i,j+1}|\,du$$

(1.4a)

$$V^2_{ij}(\hat{f}) = \frac{n}{m}\int_0^1 |(1-u)\delta_2 f_{ij} + u\delta_2 f_{i+1,j}|\,du.$$

Combining these contributions, we define the variation of the net $\hat{f}$ by

$$V(\hat{f},R) := \sqrt{2}\sum_{i=0}^{n-1}\sum_{j=0}^{m-1}\iint_{R_{ij}} |\hat{f}_{xy}|\,dxdy$$

(1.5)

$$+\sum_{i=0}^{n-1}\sum_{j=0}^{m-1} V^1_{ij}(\hat{f}) + \sum_{j=0}^{m-1}\sum_{i=0}^{n-1} V^2_{ij}(\hat{f}).$$

We now state our main

THEOREM 1. For any integers $m,n \geq 1$ and for any function f on the unit square R we have

(1.6) $$V(B_{n,m}f;R) \leq V(\hat{f}_{n,m};R).$$

2. Preliminaries for the proof.

The proof of Theorem 1 will depend upon a number of lemmas and a proposition. We begin with

LEMMA 1. Inequality (1.6) holds if $B_{n,m}f(x) \in \pi_2$.

PROOF: Choose h such that $(B_{n,m}f-h)(x,y) =, ax^2+by^2$. Then h(x,y) is quasi-linear. If g=f-h then $\hat{f}_{n,m} = \hat{g}_{n,m} + \hat{h}_{n,m}$. Moreover

$$B_{n,m}g = B_{n,m}f - h = ax^2 + by^2.$$

Since $B_{n,m}g = ax^2 + by^2$, it follows that $g_{ij} = \dfrac{ai(i-1)}{n(n-1)} + \dfrac{bj(j-1)}{m(m-1)}$ and $V(B_{mn}g;R) = 2\iint_R (a^2+b^2)^{1/2}dxdy = 2(a^2 + b^2)^{1/2}$. Using (1.4), we easily check that

$$V(\hat{g}_{n,m};R) = 2(|a| + |b|).$$

So it follows that

$$V(B_{n,m}g;R) \leq V(\hat{g}_{n,m};R).$$

Now

$$V(B_{n,m}f;R) \leq V(B_{n,m}g;R) + V(h;R)$$

$$\leq V(\hat{g}_{n,m};R) + V^{o}(\hat{h}_{n,m};R)$$

$$= V(\hat{f}_{n,m};R)$$

which completes the proof.

We now subdivide the unit square into quarters by lines $x = 1/2$ and $y = 1/2$, thus obtaining four squares $R^j (j = 1,2,3,4)$ as shown

$$\left[\begin{array}{c|c} R^4 & R^3 \\ \hline R^1 & R^2 \end{array}\right].$$

We denote the Bernstein basis function for R^r by $\beta^r_{i,j}(x,y)$. For example,

$$\beta^1_{i,j}(x,y) := \binom{n}{i}\binom{m}{j} x^i y^j \left(\frac{1}{2} - x\right)^{n-i} \left(\frac{1}{2} - y\right)^{m-i} 2^{n+m}.$$

We can now formulate

LEMMA 2. The restriction of $B_{n,m}(x,y)$ to R^r is given by

$$B_{n,m}(x,y)\Big|_{R^r} = \sum_{i=0}^{n} \sum_{j=0}^{m} a^r_{ij} \beta^r_{i,j}(x,y), \quad r=1,2,3,4 \tag{2.5}$$

where

$$\left[\begin{aligned} a^1_{ij} &= 2^{-i-j} \sum_{k=0}^{i} \sum_{l=0}^{j} \binom{i}{k}\binom{j}{l} f_{k,l} \\ a^2_{ij} &= 2^{i-j-n} \sum_{k=0}^{n-i} \sum_{l=0}^{j} f_{i+k,j+l} \binom{n-i}{k}\binom{j}{l} \\ a^3_{ij} &= 2^{-m+i+j-n} \sum_{k=0}^{n-i} \sum_{l=0}^{m-j} f_{i+k,j+l} \binom{n-i}{k}\binom{m-j}{l} \\ a^4_{ij} &= 2^{j-i-m} \sum_{k=0}^{i} \sum_{l=0}^{n-j} f_{i,j+l} \binom{i}{k}\binom{m-j}{l}. \end{aligned}\right. \tag{2.6}$$

Moreover,

$$\sum_{r=1}^{4} \sum_{i=o}^{n} \sum_{j=o}^{m} a^r_{ij} = 4 \sum_{i=o}^{n} \sum_{j=o}^{m} f_{ij}. \tag{2.7}$$

PROOF: (2.6) follows from the well-known de Casteljau subdivision alogrithm [4]. In order to prove (2.7 we observe that

$$\int_0^1\int_0^1 B_{n,m}(x,y)dxdy = \sum_{r=1}^{4} \iint_{R^r} B_{n,m}(x,y)\Big|_{R^r} dxdy.$$

The left side in the above equation equals $\frac{1}{(n+1)(m+1)} \sum_{i=0}^{n} \sum_{j=0}^{m} f_{ij}$, while the right side yields

$$\frac{1}{4(n+1)(m+1)} \sum_{i=0}^{n} \sum_{j=0}^{m} a_{ij}^{r}, \quad \text{since}$$

$$\int_0^1\int_0^1 \beta_{i,j}^{r}(x,y)dxdy = \frac{1}{4(n+1)(m+1)}.$$

Hence (2.7).

Following Goodman [4], we introduce certain averaging operators which will simplify the formula (2.6). For any array b_{ij}, we define

$$(2.8) \qquad \begin{aligned} (A_1 b)_{ij} &= \frac{1}{2}(b_{i+1,j} + b_{i-1,j}) \\ (A_2 b)_{ij} &= \frac{1}{2}(b_{i+1,j} + b_{i,j-1}). \end{aligned}$$

We then have

LEMMA 3. Let $\{b_{ij}\}_{i,j=0}^{2n,2m}$ be any array satisfying $b_{2i,2j} = f_{ij}$. Then

$$(2.9) \qquad \begin{aligned} a_{ij}^{1} &= (A_1^i A_2^j b)_{ij}, & a_{ij}^{2} &= (A_1^{n-i} A_2^j b)_{n+i,j} \\ a_{ij}^{3} &= (A_1^{n-i} A_2^{m-j} b)_{n+i,m+j}, & a_{ij}^{4} &= (A_1^i A_2^{m-j} b)_{i,m+j} \end{aligned}$$

where a_{ij}^{r} $(r = 1,2,3,4)$ are given by (2.6).

PROOF: From (2.8) it is clear that

$$(A_2^j b)_{ij} = 2^{-j} \sum_{\beta=0}^{j} \binom{j}{\beta} b_{i,2\beta}$$

so that

$$(A_1^i A_2^j b)_{ij} = 2^{-i-j} \sum_{\alpha=0}^{i} \sum_{\beta=0}^{j} \binom{i}{\alpha} \binom{j}{\beta} f_{\alpha,\beta} = a_{ij}^1,$$

by (2.6), which proves the first relation in (2.9).

Similarly, we have

$$(A_1^{n-i} A_2^j b)_{n+i,j} = \frac{1}{2^{n-i+j}} \sum_{\alpha=0}^{n-i} \sum_{\beta=0}^{j} \binom{n-i}{\alpha} \binom{j}{\beta} f_{\alpha i,\beta}$$

$$(A_1^{n-i} A_2^{m-j})_{n+i,m+j} = \frac{1}{2^{m+n-i+j}} \sum_{\alpha=0}^{n-i} \sum_{\beta=0}^{m-j} \binom{n-i}{\alpha} \binom{m-j}{\beta} f_{i+\alpha,j+\beta}$$

$$(A_1^i A_2^{m-j} b)_{i,m+j} = \frac{1}{2^{m-j+i}} \sum_{\alpha=0}^{i} \sum_{\beta=0}^{m-j} \binom{i}{\alpha} \binom{m-j}{\beta} f_{\alpha,\beta+j}$$

which completes the proof.

LEMMA 3. For any array $\{b_{ij}\}_{i=0,j=0}^{2n,2m}$, the following identity holds:

$$\sum_{i=0}^{n} \sum_{j=0}^{m} \{(A_1^i A_2^j b)_{ij} + (A_1^{n-i} A_2^j b)_{n+i,j} + (A_1^{n-i} A_2^{m-j} b)_{n+i,m+j}$$

$$(2.10) \qquad + (A_1^i A_2^{m-j} b)_{i,m+j}\} = 4 \sum_{i=0}^{n} \sum_{j=0}^{m} b_{2i,2j}.$$

3. Subdivision of the net $\hat{f}_{n,m}$.

Given $B_{n,m}f$, we have defined the net $\hat{f}_{n,m}$. Considering the restriction of $B_{n,m}f$ to $R^r (r=1,2,3,4)$, we obtain four functions g^r defined by

$$(3.1) \qquad (B_{n,m}f)|_{R^r} = B_{n,m}(g^r;R^r), \qquad r = 1,2,3,4.$$

Each of these functions g^r determines a net $\hat{g}_{nm}^r$ on R^r. By Lemma 2 we get $g_{i,j}^r$ in terms of f_{kl}. We shall prove

PROPOSITION 1. The following inequality holds:

$$(3.2)\qquad \sum_{r=1}^{4} V(\hat{g}^{r}_{n,m};R^{r}) \le V(\hat{f}_{n,m};R).$$

PROOF: Recalling the definition of $V(\hat{f}_{n,m,};R)$ we see from (1.5) that it is composed of three parts. We shall show that the above inequality holds for each of the three parts separately.

To begin with the ``continuous part``, we find by direct calculation that

$$V^{o}(\hat{f};R) = \iint_{R} |\hat{f}_{xy}|\,dxdy$$

$$(3.3)\qquad = \sum_{i=1}^{n-1}\sum_{j=0}^{m-1} |f_{i,j} - f_{i+1,j} - f_{i,j+1} + f_{i+1,j+1}|.$$

Similarly,

$$V^{o}(\hat{g}^{1},R) = \sum_{i=0}^{n-1}\sum_{j=0}^{m-1} |\hat{g}^{1}_{ij} - \hat{g}^{1}_{i+1,j} - \hat{g}^{1}_{i,j+1} + \hat{g}^{1}_{i+1,j+1}|$$

$$= \sum_{i=0}^{n-1}\sum_{j=0}^{m-1} |A_1^{i}A_2^{j}(b_{ij} - A_1 b_{i+1,j} - A_2 b_{i,j+1} + A_1A_2 b_{i+1,j+1})|$$

$$= \frac{1}{4}\sum_{i=0}^{n-1}\sum_{j=0}^{m-1} |A_1^{i}A_2^{j}(b_{i,j} - b_{i+2,j} - b_{i,j+2} + b_{i+2,j+2})|,$$

on using (2.8) and (2.9).

$$(3.4)\qquad v^{o}_{ij} := |b_{ij} - b_{i+2,j} - b_{i,j+2} + b_{i+2,j+2}|$$

we can write the above formula in a more compact form:

$$V^{o}(\hat{g}^{1},R) \le \frac{1}{4}\sum_{i=0}^{n-1}\sum_{k=0}^{m-1} A_1^{i}A_2^{j}v^{o}_{ij},$$

and similarly we obtain

$$V^{o}(\hat{g}^{2},R^{2}) \le \frac{1}{4}\sum_{i=0}^{n-1}\sum_{k=0}^{m-1} A_1^{n-i-1}A_2^{j}v^{o}_{n+i-1,j},$$

$$V^o(\hat{g}^3,R^3) \le \frac{1}{4}\sum_{i=0}^{n-1}\sum_{k=0}^{m-1} A_1^{n-i-1}A_2^{m-j-1}v^o_{n+i-1,m+j-1}$$

$$V^o(\hat{g}^4,R^4) \le \frac{1}{4}\sum_{i=0}^{n-1}\sum_{k=0}^{m-1} A_1^{i}A_2^{m-j-1}v^o_{i,m+j-1}.$$

Thus from the algebraic identity (2.10) with n-1 for n and m-1 for m, we obtain the following inequality

$$\sum_{r=1}^{4} V^o(\hat{g}^r,R^r) \le \sum_{i=0}^{n-1}\sum_{j=0}^{m-1} v^o_{2i,2j}$$

$$= \sum_{i=0}^{n-1}\sum_{j=0}^{m-1} |f_{ij} - f_{i+1,j} - f_{i,j+1} + f_{i+1,j+1}|$$

$$= V^0(\hat{f},R).$$

This proves the inequality (3.2) for the variation V^o.

We now turn to the contribution due to the discontinuity in the gradient of $\hat{f}$. From (1.4a) applied to $\hat{g}^1$, relative to the mesh on R^1, we have

$$V^1_{ij}(\hat{g}^1,R_{ij}) = \frac{n}{m}\int_0^1 |(1-u)\delta_1\hat{g}^1_{ij} + u\delta_1 g^1_{i,j+1}|\,du.$$

Since, in view of (2.9), we have

$$\delta_1\hat{g}^1_{ij} = a^1_{i-1,j} - 2a^1_{ij} + a^1_{i+1,j}$$

$$= A_1^{i-1}A_2^{j}(b_{i-1,j} - 2A_1 b_{ij} + A_1^2 b_{i+1,j})$$

$$= \frac{1}{4} A_1^{i-1}A_2^{j}(b_{i-1,j} - 2b_{i+1,j} + b_{i+3,j}),$$

it follows that

$$V^1(\hat{g}^1;R^1) = \sum_{i=0}^{n-1}\sum_{j=0}^{m-1} V^1_{ij}(\hat{g}^1,R_{ij})$$

$$\le \frac{n}{4m}\sum_{i=1}^{n-1}\sum_{j=0}^{m-1} A_1^{i-1}A_2^{j}\int_0^1 (1-u)(b_{i-1,j} - 2b_{i+1,j} + b_{i+3,j})$$

$$+ uA_2(b_{i-1,j+1} - 2b_{i+1,j+1} + b_{i+3,j+1})du$$

$$\le \frac{n}{2m}\sum_{k=0}^{n}\sum_{k=0}^{n} A_1^{i-1}A_2^{j}\int_0^{1/2} |(1-u)(b_{i-1,j} - 2b_{i+1,j} + b_{i+3,j})$$

$$+ u(b_{i-1,j+2} - 2b_{i+1,j+2} + b_{i+3,j+2})|du$$

Setting

$$V_{ij} = \int_0^{1/2} |(1-u)(b_{ij} - 2b_{i+2,j} + b_{i+4,j})$$

$$+ u(b_{i,j+2} - 2b_{i+2,j+2} + b_{i+4,j+2})|du,$$

$i = 0,\dots,2n-4$, $j = 0,\dots 2m-2$ and $v_{i,2m-1} = v_{i,2m} = 0$, we have

$$V^1(\hat{g}^1,R^1) \le \frac{n}{2m}\sum_{i=0}^{n-2}\sum_{j=0}^{m-1} A_1^iA_2^jv_{i,j}.$$

Similarly deriving corresponding formulae for $\hat{g}^2,\hat{g}^3,\hat{g}^4$, we obtain

$$V^1(\hat{g}) = \sum_{r=1}^{4} V^1(\hat{g}^r;R^r)$$

$$\le \frac{n}{2m}\sum_{i=0}^{n-2}\sum_{j=0}^{m-1}\{A_1^iA_2^jv_{ij}+A_1^{n-2-i}A_2^jv_{n-2+i,j}+A_1^{n-2-i}A_2^{m-j}v_{n-2+i,m+j}+A_1^iA_2^{m-j}v_{i,m+j}\}.$$

Applying Lemma 4 with n-2 for n and with, v_{ij} in place of b_{ij} the above yields

$$V^1(\hat{g};U) \le \frac{2n}{m}\sum_{i=0}^{n-2}\sum_{j=0}^{m} v_{2i,2j}.$$

Since $b_{2i,2j} = f_{ij}$, it follows that

$$V_{2i,2j} = \int_0^{1/2} |(1-u)(f_{i,j} - 2f_{i+1,j} + f_{i+2,j})$$

$$+ u(f_{i,j+1} - 2f_{i+1,j+1} + f_{i+2,j+1})|du$$

$$= \int_0^{1/2} |(1-u)\delta_1 f_{i+1,j} + u\delta_1 f_{i+1,j+1}| du, \quad i=0,\dots,n-1;\ j=0,\dots,m-1,$$

since $v_{2i,2m} = 0$.

Thus

$$V^1(\hat{g};U) \le \frac{2n}{m} \sum_{i=0}^{n-1} \sum_{j=0}^{m-1} \int_0^{1/2} |(1-u)\delta_1 f_{i,j} + u\delta_1 f_{i,j+1}| du.$$

The above reasoning when carried out for f(x,1-y) yields a similar inequality. In fact we obtain

$$V^1(\hat{g};U) \le \frac{2n}{m} \sum_{i=0}^{n-1} \sum_{j=0}^{m-1} \int_{1/2}^{1} |(1-u)\delta_1 f_{i,j} + u\delta_1 f_{i,j+1}| du.$$

The above two inequalities, averaged together, completes the proof of (3.2).

4. Proof of Theorem 1.

We put together Lemma 1 and Proposition 1 and follow the lines if argument used by Goodman [4] for the case of Bernstein polynomials on the triangle.

For any $p \in \pi_{n,m}$, $p = B_{n,m}(g)$ for some appropriate g on U. Then it is easy to see that both $V(g_{n,m};U)$ and $V(p;U)$ are norms on the quotient space $\pi_{n,m}/\pi_{1,1}$. As two norms on finite dimensional space are equivalent, we have

(4.1) $$V(\hat{g}_{n,m};U) \le cV(B_{n,m}g,U)$$

for some constant c. Moreover the inequality (4.1) is invariant under translation and change of scale, so that the inequality is valid for any square R.

Let $l \ge 1$ be any integer and let $\cap_l$ be the subdivision of U into l^2 smaller equal sqaures. Let σ be any one of the squares in $\cap_l$. Then

$$B_{n,m}f|_\sigma = q_\sigma + r_\sigma$$

where $q \in \pi_2$ and r_σ vanishes together with its first and second order partial derivatives about some point in σ. Then

(4.2) $$V(B_{n,m}f;\sigma) = V(q_\sigma,\sigma) + o(1/l^2).$$

Let $B_{n,m}f|_\sigma = B_{n,m}(g,\sigma)$, $q_\sigma = B_{n,m}(h,\sigma)$. Then from Lemma 1 with σ in place of U, we get

$$\begin{aligned} V(q_\sigma,\sigma) &\le V(\hat{h}_{n,m},\sigma) \\ &\le V(\hat{g}_{n,m}\sigma) + V((\hat{h}-\hat{g})_{n,m},\sigma) \\ &\le V(\hat{g}_{n,m},\sigma) + cV(B_{n,m}(h-g),\sigma). \end{aligned}$$

Since $B_{n,m}(h-g,\sigma) = q_\sigma - B_{n,m}f|_\sigma = -r_\sigma$ which is of order $o\,(1/l^2)$ in σ, we obtain

(4.3) $$V(q_\sigma,\sigma) \le V(\hat{g}_{n,m},\sigma) + o(1/l^2).$$

Combining (4.2) and (4.3), and adding over all σ in $\cap_l$, we have

(4.4) $$V(B_{n,m}f,U) \le \sum_\sigma V(\hat{g}_{n,m},\sigma) + o(1).$$

If we choose $l = 2^s$, then the proposition can be applied s times to yield

(4.5) $$V(B_{n,m}f:U) \ge V(\hat{f}_{n,m},U) + o(1).$$

Finally letting s tend to infinity, we get the result.

5. Convexity of $(B_{n,m}f)(x,y)$.

We shall now that if $(\hat{f}(x,y)$ is a convex function, then so is $B_{n,m}f$. We shall use a ``degree-raising'' argument. Such degree raising arguments have been used by Dahmen and Micchelli [3] for Bernstein polynomials on simplices and Chang and Feng [2] for triangles.

Multiplying by (x + (1 -x))(y + (1 - y) it is easy to see that

$$B_{n,m}(x,y) = \sum_{i=0}^{n} \sum_{j=0}^{m} b_{ij} \binom{n}{i} \binom{m}{j} x^i(1-x)^{n-i}y^j(1-y)^{m-j}$$

(5.1)

$$= \sum_{i=0}^{n+1} \sum_{j=0}^{m+1} b_{ij}^{*} \binom{n+1}{i} \binom{m+1}{j} x^{i}(1-x)^{n+1-i} y^{j}(1-y)^{m+1-j}$$

where

$$b_{ij}^{*} = \frac{ij}{(n+1)(m+1)} b_{i-1,j-1} + \frac{i}{n+1} (1 - \frac{j}{m+1}) b_{i-1,j}$$

(5.2)

$$+(1 - \frac{i}{n+1}) \frac{j}{m+1} b_{i,j-1} + (1 - \frac{i}{n+1})(1 - \frac{j}{m+1}) b_{ij}.$$

We will sometimes abberviate formula (5.2) as $\{b_{ij}^{*}\} = E\{b_{ij}\}$.

We want to determine the conditions for a net $\hat{f}$ to be convex. Because $\hat{f}(x,y)$ is a bilinear on R_{ij}, we have $\hat{f}_{xx} = \hat{f}_{yy} = 0$ and so we require $\hat{f}_{xy}=0$ in order to assure positive definiteness of the Hessian. This implies that

$$f_{ij} - f_{i+1,j} - f_{i,j+1} + f_{i+1,j+1} = 0,$$

(5.3)

$$i = 0,1,\ldots,n-1, \quad j = 0,1,\ldots,m-1.$$

The condition (5.3) means that the net $\hat{f}$ is linear on each patch R_{ij}. It is then geometrically clear that convexity of $\hat{f}$ will follow from (5.3) and the following two conditions:

(5.4) $$f_{i+1,j} - 2f_{i,j} + f_{i-1,j} \geq 0, \quad i = 1,\ldots,n-1, \quad j = 0,\ldots,m$$

(5.5) $$f_{i,j+1} - 2f_{ij} + f_{i,j+1} \geq 0 \qquad j = 1,\ldots,m-1, \quad i = 0,\ldots,n.$$

We can now state

THEOREM 2. If $\hat{f}_{n,m}$ is convex on R, then so is $B_{n,m}f$. More precisely, (5.3), (5.4) and (5.5) together constitute a sufficient condition for the convexity of $B_{n,m}f$.

PROOF: We shall show that the three conditions (5.3) - (5.5) are all preserved under the degree-raising transformation $\{f_{ij}^{*}\} = E\{f_{ij}\}$. Using (5.2) we have

$$(m+1)(n+1)f_{ij}^{*} = ijf_{i-1,j-1} + i(m+1-j)f_{i-1,j}$$

(5.6)

$$+ (n+1-i)jf_{i,j-1} + (n+1-i)(m+1-j)f_{ij}.$$

If in (5.3) we replace i by i - 1 and j by j - 1, we can get $f_{i-1,j-1}$ in terms of $f_{i,j-1}, f_{i-1,j}$ and f_{ij}. Then (5.6), after simplifying, becomes

$$(m+1)(n+1)f^*_{ij} = j(n+1)f_{i,j-1} + i(m+1)f_{i-1,j} + \{(i+1)(m+1) - j(n+i)\}f_{ij}.$$

Similarly we get the following formulae:

$$(m+1)(n+1)f^*_{i+1,j} = (n+1)(jf_{i,j-1} + \{(i+1)(m+1) - j(n+1)\}f_{ij} + (n-i)(m+1)f_{i+1,j}$$

$$(m+1)(n+1)f^*_{i,j+1} = i(m+1)\, f_{i-1,j} + \{(n+1)(j+1) - i(m+1)\}f_{ij} + (m-j)(n+1)f_{i,j+1}.$$

$$(m+1)(n+1)f^*_{i+1,j+1} = \{(i+1)(j+1) - (n-i)(m-j)\}f_{ij} + (m-j)(n+1)f_{i,j+1} + (n-i)(m+1)f_{i+1,j}.$$

Now we can compute

(5.7) $$f^*_{ij} - f^*_{i+1,j} - f^*_{i,j+1} + f^*_{i+1,j+1}.$$

Thus if we collect the coefficient of $f_{i,j-1}$, $f_{i-1,j}$, $f_{i,j}$, $f_{i,j+1}$ and $f_{i+1,j}$, we see that they all vanish which proves the assertion that (5.3) is preserved under E.

It remains to show that relations (5.4) and (5.5) are preserved by the degree raising operation. Employing (5.6), with i replaced by i+1 and i-1 as necessary, we obtain

$$\begin{aligned}
&(m+1)(n+1)\{f^*_{i+1,j} - 2f^*_{ij} + f^*_{i-1,j}\} \\
&= \{(i+1)f_{i,j-1} - 2if_{i-1,j-1} + (i-1)f_{i-2,j-1}\}j \\
&\quad + (m+1-j)\{(i+1)f_{ij} - 2if_{i-1,j} + (i-1)f_{i-2,j}\} \\
&\quad + j\{(n-i)f_{i+1,j-1} - 2(n+1-i)f_{i,j-1} + (n+2-i)f_{i-1,j}\} \\
&\quad + (m+1-j)\{(n-i)f_{i+1,j} - 2(n+1-i)f_{ij} + (n+2-i)f_{i-1,j}\} \\
&\geq 2j(f_{i,j-1} - f_{i-1,j-1}) + 2(m+1-j)(f_{ij} - f_{i-1,j})
\end{aligned}$$

$$+ 2j(f_{i+1,j-1} - f_{i,j-1}) + 2(m+1-j)(f_{i-1,j} - f_{ij})$$

$$= 0.$$

Similarly, the inequality (5.5) is preserved under the degree raising operation.

By iterating the degree raising operation, we obtain

$$\{g^p_{ij}\}^{n+p,m+p}_{i,j=1} = E^p\{f_{ij}\}, \quad p = 1,2,\dots .$$

Then it follows that the net $\hat{g}_p$ determined by the values g^p_{ij} is convex. Since

$$\lim_{p \to \infty} || \hat{g}_p - B_{n,m}f ||_\infty = 0$$

we conclude that $B_{n,m}f$ is convex.

An example to demostrate the role of (5.4) is easily given. Take n = m = 2 and let

$$f_{ij} = 0, \quad i = 0,1, \quad j = 0,1, \quad f_{22} = 1.$$

Then (5.4) and (5.5) are fulfilled. But $B_{22}(x,Y) = x^2y^2$ which is not convex. So (5.4) and (5.5) alone are not sufficient to give convexity. Furthermore the equality in (5.3) cannot be weakened to an ineqaulity.

REFERENCES

1. Geng-zhe Chang and Philip J. Davis, The convexity of Bernstein polynomials over triangles, J. Approximation Theory 40 (1984), 11-28.

2. Geng-zhe Chang and Yu-yu Feng, A new proof for convexity of Bernstein polynomials over triangles, to appear in Chinese Journal of Mathematics, Series B.

3. W. Dahmen and C.A. Micchelli, Convexity of multivariate Bernstein polynomials and box spline surfaces, Studia mathematicarum Hungarica 23(1988), 265-287.

4. G. Farin, Curves and Surfaces for Computer aided Geometic Design, Academic Press, N.Y., 1988.

5. T.N.T. Foodman, Variation diminishing properties of Bernstein polynomials on triangles, J. Approximation Theory 50 (1987), 111-126.

Department of Mathematics
University of Alberta
Edmonton, Alberta T6G 2G1

Department of Mathematics
Kent State University
Kent, Ohio, U.S.A. 44214

APPROXIMATION IN WEIGHTED BANACH SPACES OF POWER SERIES

B.L.Wadhwa and B.S.Yadav

Let $l^1\{w\}$ be a weighted Banach space of power series with the weight sequence $w = \{w_n\}_{n=0}^{\infty}$. Also, let S denote the left(right) shift operator on $l^1\{w\}$. This paper is concerned with the investigation of conditions on a vector $x \in l^1(w)$ and on the weight sequence $\{w_n\}$ under which

$$\overline{span}\, \{S^n x,\ n = 0,1,\dots\dots\} = l^1(w).$$

In case $l^1(w)$ is a Banach algebra, it is shown that these approximation problems are related with the algebra of multipliers of a maximal ideal of $l^1(w)$.

§1. We consider the algebra $\mathbb{C}[[z]]$ of formal power series

$$(1.1) \qquad x = x(z) = \sum_{n=0}^{\infty} a_n z^n ,$$

where the coefficients a_n's are complex. For a sequence $w = \{w_n\}_{n=0}^{\infty}$ of positive numbers and we denote by $l^1(w)$ the Banach space of all vectors $x \in \mathbb{C}[[z]]$ with

---

This work was done while the second author was visiting Cleveland State University (CSU) on sabbatical leave from the University of Delhi. He wishes to thank the Department of mathematics,CSU,for their help and kind hospitality.

$$||x|| = \sum_{n=0}^{\infty} |a_n| w_n < \infty .$$

We shall identify an element $x \in l^1(w)$ given by (1.1) with the sequence

(1.2) $$a = \{a_n\}_{n=0}^{\infty}$$

and the Banach space $l^1(w)$ with the Banach space of all complex sequences (1.2) for which the norm is defined by

$$||a|| = \sum_{n=0}^{\infty} |a_n| w_n .$$

The left shift L and the right shift R on $l^1(w)$ in that case appear as:

(1.3) $$La = \{a_1, a_2, a_3, \ldots\ldots\},$$

(1.4) $$Ra = \{0, a_0, a_1, \ldots\ldots\}.$$

The object of this paper is to investigate the following approximation problems: For which vectors $x \in l^1(w)$, under suitable conditions on the weight sequence $\{w_n\}$, one has

(1.5) $$\bigvee_{n=0}^{\infty} \{L^n x\} = l^1(w),$$

and a similar question for the right shift R, where the left hand side of (1.5) denotes $\overline{span}\{L^n x : n = 0,1,2,\ldots.\}$. Since the closed ideal generated by an element x in $l^1(w)$ coincides with the closure of the subspace generated by the vectors $R^n x$, $n = 0.1.2\ldots,$ these approximation problems are intimately connected with those of characterizing the standard closed ideals or showing the existence of non-standard closed ideals in the socalled radical Banach algebras $\iota^1(w)$; see Thomas [4]-[6].

In the end we shall show that, in the case of Banach algebras $\iota^1(w)$, these approximation problems are related to the algebra of multipliers of a maximal ideal of $\iota^1(w)$ which has been studied by Bade, Dales and Laursen [1].

$\xi 2.$ Let $\sigma = \{\sigma_n\}_{n=o}^{\infty}$ be a bounded sequence of positive numbers. The weighted left shift L_σ on the Banach space ι^1 of all absolutely summable complex sequences $\alpha = \{\alpha_n\}_{n=o}^{\infty}$ is defined by

$$L_\sigma \alpha = \{\sigma_0 \alpha_1, \sigma_1 \alpha_2, \sigma_2 \alpha_3, \ldots\}$$

and the weighted right shift R_σ by

$$R_\sigma \alpha = \{0, \sigma_0 \alpha_0, \sigma_1 \alpha_1, \ldots\}.$$

The (unweighted) left shift L (right shift R) on the weighted Banach space $\iota^1(w)$ can be studied via the weighted left shift L_σ (the right shift R_σ) on ι^1 and vice versa, provided that

(2.1) $w_o = 1$ and $w_n = \sigma_0 \sigma_1 \ldots \ldots \sigma_{n-1}$ for $n \geq 1$.

We make use of this observation in proving our first result for Banach spaces $\iota^1(w)$.

THEOREM 1. Let $\times$ be a vector in the Banach space $\iota^1(w)$ with infinitely many $\alpha_n \neq 0$, and let the weight sequence $\{w_n\}$ satisfy the following conditions:

(2.2) $$\delta = \sup_{k,n\geq N} \left\{ \frac{w_{k+n-N}}{w_k \cdot w_n} \right\} < \infty$$

for all sufficiently large positive integers N,

(2.3) $$\sum_{j=N}^{\infty} \frac{(j+1)w_j}{w_{j-N}} < \infty$$

for some large positive integer N. Then (1.5) holds.

We need the following lemma which is implicit in the proof of the main theorem in Nikolskiǐ [3].

LEMMA. Let E be a closed subspace of ℓ^1. If for every vector $\alpha = \{\alpha_n\}_{n=0}^{\infty}$ in E,

(2.4) $$|\alpha_n| \leq \beta_n ||\alpha||, \quad \sum \beta_n < \infty,$$

where the constants β_n are independent of α, then E is finite-dimensional. The same result holds for E^{o}, the annihilator of E, in ℓ^{∞}.

PROOF. We shall show that the closed unit ball B of E is compact and hence E is finite-dimensional by Riesz theorem. Let $\{e_n\}_{n=0}^{\infty}$ denote the standard basis of ℓ^1. For a given $\varepsilon > 0$, there exists an integer $p \geq 0$ by (2.4) such that

(2.5) $$\sum_{n=p+1}^{\infty} \beta_n < \varepsilon/2.$$

Let

$$E_p = \bigvee_{n=0}^{p} \{e_n\}.$$

Then for each vector $\alpha = \{\alpha_n\}_{n=0}^{\infty}$ in B, there is a vector (by (2.4) and (2.5))

$$\alpha^{(p)} = \sum_{n=0}^{p} \alpha_n e_n$$

in the closed unit ball B_p of the finite-dimensional subspace E_p such that

(2.6) $$\| \alpha^{(p)} - \alpha \| < \varepsilon/2 .$$

Since B_p is compact, it has an $\varepsilon/2$-net τ. It follows from (2.6) that this τ serves as an ε-net of B. Thus B is complete and totally bounded, and hence compact.

For the proof of the theorem, we put

$$E = \bigvee_{n=0}^{\infty} \{L_\sigma^n \alpha\} .$$

It will suffice to show that $E = \iota^1$ under the conditions (2.2) and (2.3). Suppose that $E \neq \iota^1$. Then there is a non-zero $b = \{b_o, b_1 \ldots\}$ in ι^∞ such that $b(E) = o$, or equivalently $b(L_\sigma^n \alpha\} = o$ for all $n = o, 1, \ldots$. Let $m = \min \{k : b_k \neq o\}$. Since

$$L_\sigma^n \alpha = \{\sigma_0 \sigma_1 \ldots \sigma_{n-1} \alpha_n, \sigma_1 \sigma_2 \ldots \sigma_n \alpha_{n+1}, \ldots\},$$

we have

$$0 = b(L_\sigma^n \alpha) = \sum_{k=o}^{\infty} \sigma_k \sigma_{k+1} \ldots \sigma_{k+n-1} \alpha_{k+n} b_k$$

$$= \sum_{k=m}^{\infty} \frac{w_{k+n}}{w_k} a_{k+n} b_k , \qquad \text{(by (2.1))}$$

and hence, as $b_m \neq o$,

$$-a_{m+n} = \sum_{k=m+1}^{\infty} \frac{w_m \; w_{k+n}}{w_{m+n} \; w_k b_m} \cdot a_{k+n} b_k .$$

Now choosing N as in condition (2.3), we can choose b such that $b_k = o$, for $m < k < m + n$. In that case

$$|a_{m+n}| \leq C \sum_{k=m+N}^{\infty} \frac{w_{k+n}}{w_{m+n} \; w_k} |a_{k+n}| \, ||b|| ,$$

where C denotes a constant depending on m.

Since, by condition (2.2),

$$\frac{w_{k+n}}{w_{m+n} \cdot w_k} = \frac{w_{k+n-n}}{w_{m+n} \cdot w_k} \cdot \frac{w_{k+n}}{w_{k+n-n}} \leq \delta \frac{w_{k+n}}{w_{k+n-N}} ,$$

we have

$$|a_{m+n}| \leq \left(C||b||\delta \sum_{k=m+N}^{\infty} \frac{w_{k+n}}{w_{k+n-N}} \right) ||a|| .$$

Now, we observe that the vector a can be replaced by any vector $f = \{f_n\}_{n=o}^{\infty}$ in E, the space spanned by a, in the above calculations. Thus we have, for all $f \in E$,

$$|f_{m+n}| \leq \beta_{m+n} \, ||f|| ,$$

where the contants

$$\beta_{m+n} = C||b||\delta \sum_{k=m+N}^{\infty} \frac{w_{k+n}}{w_{k+n-N}}$$

do not depend upon the choice of f in E. Taking $\beta_i = 1$ for $i \neq m+n$, the condition (2.3) ensures that

$$\sum_{n=1}^{\infty} \beta_n < \infty.$$

Now it follows by the lemma that E is finite-dimensional. But this contradicts our assumption that $\alpha_n \neq 0$ for infinitely many n. Therefore $E = \iota^1$.

REMARK. As an example to illustrate our theorem, we see that the condition (2.2) holds when $\{\sigma_n\}$ is monotonically decreasing, or more generally, when $\{\sigma_n\}$ is monotonically decreasing on certain arithmetic progressions (e.g. when $\{\sigma_{kr+j}\}$ is monotonically decreasing as $k \to \infty$ and $j = 0, \ldots\ldots, r-1$). Also, if there exists an $\varepsilon > 0$ such that

$$\prod_{j=0}^{r-1} \sigma_{kr+j} = 0\,(r^{-\varepsilon}) \text{ as } r \to \infty,$$

then the condition (2.3) holds.

We now assume that the sequence $\{w_n\}$ further satisfies the following conditions:
i) $w_0 = 1$, ii) $0 < w_n < 1$ for all $n > 0$ and (iii) $w_{m+n} \leq w_m w_n$ for all m,n and define multiplication in $\iota^1(w)$ by convolution: For any $x,y \in \iota^1(w)$, where x is given by (1.1) and

$$y = \sum_{n=0}^{\infty} b_n z^n ,$$

the convolution product $x * y$ is defined by

$$x * y = \sum_{n=0}^{\infty} c_n z^n ,$$

where

$$c_n = \sum_{k=0}^{n} \alpha_k b_{n-k}.$$

Thus $\ell^1(w)$ is a commutative Banach algebra. Such algebras of power series have been studied by a number of authors; see for example, references [1],[2],[4]-[6].

Our next theorem shows that (1.5) holds in every Banach algebra $\ell^1(w)$ without additional conditions on the weight sequence $\{w_n\}$ provided that the coefficients α_n's of a vector x satisfy certain growth conditions.

THEOREM 2. Let x be a vector in the Banach algebra $\ell^1(w)$ with infinitely many $\alpha_n \neq 0$ such that

$$(2.7) \qquad \lim_{n\to\infty} \sum_{k=n+1}^{\infty} \left|\frac{\alpha_k}{\alpha_n}\right| = 0,$$

then (1.5) holds.

PROOF. Let

$$E = \bigvee_{n=0}^{\infty} \{L_\sigma^n \alpha\}.$$

It will suffice to show that, under (2.7), $E = \ell^1$. Considering the weighted shift L_σ on ℓ^1, we have

$$L_\sigma^n \alpha = \{\sigma_0 \sigma_1 \ldots \sigma_{n-1} \alpha_n ,\ \sigma_1 . \sigma_2 \ldots \sigma_n \alpha_{n+1} ,\ \ldots\}.$$

Now for $\alpha_n \neq 0$,

$$\left\| \frac{L_\sigma^n \alpha}{\sigma_0\sigma_1\cdots\sigma_{n-1}\alpha_n} - e_0 \right\| = \left\| \sum_{k=n+1}^{\infty} \frac{\sigma_{k-n}\ \cdots\ \sigma_{k-1}\alpha_k}{\sigma_0\sigma_1\cdots\sigma_{n-1}\alpha_n} e_{k-n} \right\|$$

$$\leq \sum_{k=n+1}^{\infty} \frac{\sigma_0\sigma_1\ \cdots\ \sigma_{k-n}\cdots\ \sigma_{k-1}}{\sigma_0\sigma_1\cdots\sigma_{k-n-1}\cdot\sigma_0\sigma_1\cdots\sigma_{n-1}} \left|\frac{\alpha_k}{\alpha_n}\right|$$

$$= \sum_{k=n+1}^{\infty} \frac{w_k}{w_{k-n}\cdot w_n} \left|\frac{\alpha_k}{\alpha_n}\right| \quad \text{(by (2.1))}$$

$$\leq \sum_{k=n+1}^{\infty} \left|\frac{\alpha_k}{\alpha_n}\right| \quad \text{(by (1.4))}$$

$$\to 0, \text{ as } n \to \infty . \quad \text{(by(2.7))}$$

This shows that $e_0 \in E$, and hence for each n, so does

$$L_\sigma^n \alpha - \sigma_0\sigma_1\cdots,\sigma_{n-1}\alpha_n e_0,$$

or

$$\sum_{k=n+1}^{\infty} \sigma_{k-n}\cdots\sigma_{k-1}\alpha_k e_{k-n}.$$

Thus whenever $\alpha_{n+1} \neq 0$,

$$e_1 + \sum_{k=n+2}^{\infty} \frac{\sigma_{k-n}\ \cdots\ \sigma_{k-1}\alpha_k}{\sigma_1\sigma_2\ \cdots\ \sigma_n\alpha_{n+1}} e_{k-n}$$

is in E. Now proceeding as in the case of e_0 above, we can show that $e_1 \in E$. It is clear that $e_n \in E$ for all n by induction, and hence $E = \iota^1$.

§3. In this section, we prove an approximation theorem for the right shift R on the Banach space $\iota^1(w_n)$.

THEOREM 3. If the weight sequence $\{w_n\}$ satisfies the conditions:

(3.1) $$w_{k+n+1} \leq C\, w_{k+1} \cdot w_{n+1}$$

(3.2) $$\sum_{n=0}^{\infty} \frac{w_{n+1}}{w_n} < \infty ,$$

then for any vector x with $\alpha_0 \neq 0$ in the Banach space $\iota^1(w)$,

(3.3) $$\bigvee_{n=0}^{\infty} \{R^n x\} = \iota^1(w).$$

PROOF. We put

$$E = \bigvee_{n=0}^{\infty} \{R^n \alpha\}$$

and, in order to prove (3.3), we show that under the condition (3.1) and (3.2), $E = \iota^1$. Let $b = \{b_0, b_1, \ldots..\}$ be any vector in E^0, the annihilator of E, in ι^∞ such that $b(R_\sigma^n \alpha) = o$ for all $n = 0,1,2, \ldots..$ It will suffice to show that b = 0. As

$$R_\sigma^n \alpha = \left\{ o, o, \ldots, o, \sigma_o \sigma_1 \ldots \sigma_{n-1} \alpha_o , \sigma_1 \sigma_2 \ldots \sigma_n \alpha_1 , \ldots \right\},$$

$b(R_\sigma^n \alpha) = 0$ implies that

$$-b_n = \sum_{k=o}^{\infty} \frac{\sigma_{k+1}\sigma_{k+2}\ldots\sigma_{k+n}}{\sigma_o\sigma_1\ldots\sigma_{n-1}\alpha_o} \alpha_{k+1} b_{k+n+1}$$

$$= \frac{\sigma_n}{\alpha_o} \sum_{k=o}^{\infty} \frac{w_{k+n+1}}{w_{k+1}\cdot w_{n+1}} \alpha_{k+1} b_{k+n+1}, \quad (\text{by}(2.1))$$

and hence

$$|b_n| \le \frac{\sigma_n}{|\alpha_o|} ||b|| \sum_{k=o}^{\infty} \frac{w_{k+n+1}}{w_{k+1}\cdot w_{n+1}} |\alpha_{k+1}|$$

$$\le \frac{C\sigma_n}{|\alpha_o|} ||b|| \sum_{k=o}^{\infty} |\alpha_{k+1}| \quad (\text{by}(3.1))$$

$$\le \frac{C\sigma_n}{|\alpha_o|} ||b|| \, ||\alpha||.$$

Now we see that

$$\sum_{n=0}^{\infty} \sigma_n = \sum_{n=0}^{\infty} \frac{w_{n+1}}{w_n} < \infty$$

by (3.2). Therefore it follows by the lemma that E^o is finite-dimensional. Since R_σ is quasinilpotent, so is its adjoint R_σ^*; and hence $R_\sigma^*|_{E^o}$ is nilpotent. Thus there exists an integer $m \ge 0$ such $(R_\sigma^*)^m = 0$ and $(R_\sigma^*)^{m-1} \ne 0$ on E^o. As R_σ^* is the weighted left shift with weights $\{\sigma_n\}$ on ι^∞, we have

$$(R_\sigma^*)^m (b) = \{\sigma_0\sigma_1\ldots\sigma_{m-1}b_m, \sigma_1\sigma_2\ldots\sigma_m b_{m+1}, \ldots\}.$$

This shows that $b_k = 0$ for all $k \geq m$, and consequently for every $c = \{c_k\}$ in E, $c_k = 0$ for all $k < m$. Since $\alpha \in E$ with $\alpha \neq 0$, it follows that $m = 0$. Thus $b_k = 0$ for all k and $b = 0$. This completes the proof.

REMARK. It is easy to see that the condition (3.1) is more general than the condition that $\{\sigma_n\}$ be monotonically decreasing. Moreover, the following example shows that the conditions (3.1) and (3.2), put together, are independent of this condition: Let $\{\sigma_n\}$ be such that

$$w_0 = w_1 = 1,$$

$$w_{2n} = 1/3^2.2^2.5^2.4^2\ldots.(2n+1)^2,$$

and $$w_{2n+1} = 1/3^2.2^2.5^2.4^2\ldots.(2n+1)^2(2n)^2,$$

where $n = 1,2,3\ \ldots.$ Then we see that $\{w_n\}$ satisfies the condition (3.1). Also, as w_{n+1}/w_n is equal to $1/n^2$ or $1/(n+2)^2$ according to n is even or odd, the condition (3.2) is also satisfied. However, the sequence $\{\sigma_n\} = \{w_{n+1}/w_n\}$ is obviously not monotonically decreasing. This observation acquires importance in the light of the recent work of Yakubovich [7].

Let A be a commutive Banach algebra and let T be a linear operator on A such that $T(xy) = xTy$, for all $x,y \in A$. Then T is called a multiplier of A. If $B(A)$ denotes the Banach algebra of all bounded linear operators on A and $M(A)$ the algebra of all continuous multipliers of A, then $M(A)$ is a strongly closed subalgebra of $B(A)$. Now, for each $x \in A$, define an operator T_x on A by $T_x(y) = xy$ for all $x,y \in A$. Then $T_x \in M(A)$ and the mapping $x \rightarrow T_x$ is called the regular representation of A. This shows that A can be regarded, via its regular representation, as an ideal in $M(A)$.

Here we are concerned with the Banach algebra $\iota^1(w)$ and its maximal ideal

$$M = \{ x \in \iota^1(w) : x(0)=0 \}.$$

Multilpliers of algebras $\iota^1(w)$ are studied in [1]. In particular, it is shown there that L(M), the image of M under the left shift L, coincides with $M(M)$ if and only if L(M) is a convolution Banach algebra. Here we obtain the following result involving $M(M)$.

COROLLARY. If $\iota^1(w)$ is a Banach algebra of power series such that L(M) $= M(M)$ and (3.2) is satisfied, then (3.3) holds for every $x \in \iota^1(w)$ with $\alpha_0 \neq 0$.

PROOF. Since L(M) = $M(M)$, it has been shown in [1] that there exists a constant $C > 0$ such that

$$w_{m+n+1} \leq C\, w_{m+1} \cdot w_{n+1} \quad \text{for all } m,n ,$$

the condition (3.1) in Theorem 3.

REFERENCES.

[1] W.G.Bade, H.G.Dales and K.B.Laursen, Multipliers of radical Banach algebras of power series, Memoires Amer. Math.Soc.,49(1984), No. 303.

[2] S.Grabiner, Weighted shifts and Banach algebras of power series, Amer.J.Math.,97(1975), 17-42.

[3] N.K.Nikolskii, Invariant subspaces of certain completely continuous operators, Vestnik Leningrad Univ.(7), 20(1965), 68-77.(Russian)

[4] Marc P.Thomas, Closed ideals and biorthogonal systems in radical Banach algebras of power series, Proc. Edinburgh Math. Soc.,25(1982), 245-257.

[5] Marc P.Thomas, Closed ideals of $\iota^1(w)$ when $\{w_n\}$ is star-shaped, Pacific J.Math., 105(1983),237-255.

[6] Marc P.Thomas, A non-standard ideal of a radical Banach algebra of power series, Acta Math., 152(1984), 199-217.

[7] D.V.Yakubovich, Invariant subspaces of weighted shift operators, Zapiski Nauk Sem.LOMI, 141(1985), 100-143 (Russian) = J.Soviet Math., 37(1987), 1323-1346.

Cleveland State University
Cleveland, Ohio 44115 USA
and
University of Delhi
Delhi - 110007,INDIA.

A Note On Generalised Commutativity Theorems In The Schatten Norm

B.P.DUGGAL

Dedicated to the memory of U.N. Singh

1. Let H be a separable infinite dimensional complex Hilbert space, and let B(H) denote the algebra of operators (i.e., bounded linear transformation) on H into itself. Let C_p $(=C_p(H))$, $1\leq p<\infty$, denote the space $\{A\in B(H):|A|^p\in C_1\}$, where $C_1=C_1(H)$ denotes the space of trace class operators and $|A|=(A^*A)^{1/2}$, with norm $||A||^p=(tr|A|^p)^{1/p}$. The spaces C_p are Banach spaces. The well known Putnam-Fuglede commutativity theorem says that if A and B are normal operators, then AX-XB=0 implies $A^*X-XB^*=0$ for all $X\in B(H)$. Extending the Putnam-Fuglede theorem to modulo the Hilbert-Schmidt class C_2, Weiss [6,7] has shown that for given normal operators A and B, and $X\in B(H)$, $||AX-XB||_2=||A^*X-XB^*||_2$. A natural generalisation of this result is obtained upon considering normal operators A_j and B_j, j=1,2 such that $[A_1,A_2]=[B_1,B_2]=0$ (where for $A,B\in B(H)$, [A,B]=AB-BA). One has in such a case that $||A_1XB_1+A_2XB_2||_2=||A_1XB_2||_2$ ([6, Corollary 2 of part II]; see also [5] and [8], where the case of n such mutually commuting operators is considered).

The space C_p is said to possessthe GPFP (=generalised Putnam-Fuglede property) if for any normal operators A and B, and $X\in B(H)$, we have $||AX-XB||_p<\infty$ if and only if $||A^*X-XB^*||_p<\infty$. One proves the GPFP for spaces C_p by providing an estimate of the form

(1) $$||AX-XB||_p \leq d_p \ ||A^*X-XB^*||_p,$$

where d_p is a constant depending only on p. It is clear from the results of Weiss that C_2 has GPFP (with $d_p=1$). If $p\neq 2$, (1) can hold with $d_p=1$ for all $X\in B(H)$ if and only if the normal operators A and B have their spectrum on a straight line or on a circle [4]. In a recent paper [1], Abdessemed and Davies have shown that every C_p space with $2 \le p < \infty$ has the GPFP.

In this note we combine a technique due to Apostol (see [6, part II, pp 12-14])
and the GPFP result of Abdessemed and Davies [1, Corollary 3.5] to prove:

THEOREM 1. Let A_j and B_j, j=1,2, be normal operators satisfying $[A_1,A_2]=[B_1,B_2]=0$. If $2 \le p<\infty$ and $\sum_{j=1}^{2} A_j^*XB_j^*\in C_p$, then

$$\text{(2)} \qquad \|\sum_{j=1}^{2} A_j XB_j\|_p \le d_p\|\sum_{j=1}^{2} A_j^*XB_j^*\|_p$$

for all $X\in B(H)$ and for some constant d_p (≥ 1) dependent only on p.

2. Proof of Theorem 1. Defining $T_j = A_j \oplus B_j$, j =1,2, and $Y=\begin{bmatrix} 0 & X \\ 0 & 0 \end{bmatrix}$, on $\hat{H} = H \oplus H$, it is seen that (2) holds if and only if

$$\text{(3)} \qquad \|\sum_{j=1}^{2} T_j YT_j\|_p \le d_p\|\sum_{j=1}^{2} T_j^*YT_j^*\|_p,$$

where T_j are normal, $[T_1,T_2]=0$, $Y\in B(\hat{H})$ and $\sum_{j=1}^{2} T_j^*YT_j^*\in C_p(\hat{H})$. Let kerA denote the kernel of the operator A. Decompose T_1, with respect to the direct sum decomposition $\hat{H}=(\ker T_1)^{\perp}\oplus \ker T_1$ by $T_1=E_1+0$, and let

Y have the matrix representation $Y=\begin{bmatrix} Y_1 & Y_3 \\ Y_4 & Y_2 \end{bmatrix}$ with respect to this decomposition of H. Since $[T_1,T_2]=0$ implies $[T_1,T_2^*]=0$ (by the Putnam-Fuglede theorem), $(\ker T_1)^{\perp}$ reduces T_2, and so $T_2=F_1\oplus F_2$ (where $F_1=T_2|(\ker T_1)^{\perp}$). Then

$$(4) \qquad \sum_{j=1}^{2} T_jYT_j = \begin{bmatrix} E_1Y_1E_1+F_1Y_1F_1 & F_1Y_3F_2 \\ F_2Y_4F_1 & F_2Y_2F_2 \end{bmatrix}$$

Now decompose F_1, with respect to $(\ker T_1)^{\perp}=(\ker F_1)^{\perp}\oplus F_1$, by $F_1 \oplus 0$, and let Y_1 have the matrix representation $Y_1 = \begin{bmatrix} Y_{11} & Y_{13} \\ Y_{14} & Y_{12} \end{bmatrix}$

Since $[E_1,F_1]=0$, $E_1=E^1 \oplus E^2$ with respect to this decomposition of $(\ker A_1)^{\perp}$, and we have

$$E_1Y_1E_1 + F_1Y_1F_1= \begin{bmatrix} E^1Y_{11}E^1+F^1Y_{11}F^1 & E^1Y_{13}E^2 \\ E^2Y_{14}E^1 & E^2Y_{12}E^2 \end{bmatrix}$$

Defining E_{12}, E_{21}, Y_{01} and Y_{10} by $E_{12}=E^1\oplus E^2$ $(=E_1)$, $E_{21}=E^2\oplus E^1$, $Y_{01}=\begin{bmatrix} 0 & Y_{13} \\ 0 & 0 \end{bmatrix}$ and $Y_{10} = \begin{bmatrix} 0 & Y_{14} \\ 0 & 0 \end{bmatrix}$, it is now seen that

$$\|E_1Y_1E_1+F_1Y_1F_1\|_p=\|E^1Y_{11}E^1+F^1Y_{11}F^1\|_p + \|E_{12}Y_{01}E_{12}\|_p + \|E_{21}Y_{10}E_{21}\|_p + \|E^2Y_{14}E^2\|_p,$$

where E^1, F^1, E^2, E_{12} and E_{21} are all injective normal operators and $[E^1,F^1]=0$. (Here by a slight misuse of notation we have used $\|\cdot\|_p$ to denote C_p-norms with respect to different underlying separable Hilbert spaces.) A similar argument shows that the C_p-norms of the remaining entries in the matrix (4) can be written so that the normal operators appearing therein are all injective.

Also $\|\sum_{j=1}^{2} T_j^* Y T_j^*\|_p$ results in a precisely corresponding sum of the C_p-norms of the operators of the type $\sum_{j=1}^{2} E_j^* Z E_j^*$ ($\in C_p$), so that it is sufficient to consider the case of normal injective operators T_j $(j=1,2)$.

The next step in the argument is to show that we may assume the injective normal operators T_j to be invertible. Since $[T_1,T_2]=0$, the spectral theorem implies the existence of operators M_ϕ representing T_j, where $\phi_j \in L^\infty(T)$, M_ϕ (acting on $L^\infty(T)$) is the $_j$ operator of multiplication by ϕ_j and the Lebesgue measure of $\{z:\phi_j(z)=c\}$ equals zero for every complex number c. The projections P_m determined by

$$\{|\phi_1| > 1/m\} \cap \{|\phi_2| > 1/m\}$$

commute with T_j and are such that $P_m T_j \to T_j$ uniformly and $P_m T_j$ is bounded below on $(\ker P_m T_j)^\perp$. Let Q_m denote the orthogonal projections onto

$$L^2(\{|\phi_1| \le 1/m\} \cup \{|\phi_2| \le 1/m\}).$$

Then $P_m Q_m=0$, and $(P_m T_j+(1/m)Q_m)$ are invertible normal operators such that

$$[P_m T_1+(1/m)Q_m,\ P_m T_2 + (1/m)Q_m]=0$$

and

$$P_m\Big(\sum_{j=1}^{2} T_j Y T_j\Big)P_m = \sum_{j=1}^{2} (P_m T_j + (1/m)Q_m)\, P_m Y_m\, (P_m T_j+(1/m)Q_m).$$

Notice that if T is an invertible normal operator then $||TYT||_p$ $= ||UT^*YT^*U||_p$, where $U = TT^{*-1}$ is unitary. Hence $||TYT||_p = ||T^*YT^*||_p$.

If T_1 and T_2 are invertible normal operators such that $[T_1,T_2]=0$, then

$$||\sum_{j=1}^{2} T_jYT_j||_p = ||U_1(T_1^*YT_1T_2^{-1}T_2^*+T_1^{-1}T_2T_1^*YT_2^*)U_2||_p,$$

where $U_1=T_1T_1^{*-1}$ and $U_2=T_2T_2^{*-1}$ are unitary. Hence

$$||\sum_{j=1}^{2} T_jYT_j||_p = ||ZT_1^*T_2^{*-1}+T_1^{*-1}T_2^*Z||_p,$$

where $Z=T_1^*YT_2^*$, and $T_1T_2^{-1}, T_1^{-1}T_2$ are normal operators. Since $\sum_{j=1}^{2} T_j^*YT_j^* \in C_p$, we have from [1, Corollary 3.5] that there exists a scalar c_p (dependent only on p) such that

$$||\sum_{j=1}^{2} T_jYT_j||_p \leq c_p||ZT_1^*T_2^{*-1} + T_1^{*-1}T_2^*Z||_p$$
$$=c_p ||\sum_{j=1}^{2} T_j^*YT_j^*||_p.$$

Finally, we notice that

$$||P_m(\sum_{j=1}^{2} T_jYT_j)P_m||_p =||\sum_{j=1}^{2} (P_mT_j + (1/m)Q_m)P_mYP_m(P_mT_j+(1/m)Q_m)||_p$$

$$\leq c_p ||\sum_{j=1}^{2} (P_mT_j+(1/m)Q_m)^* P_mYP_m(P_mT_j+(1/m)Q_m)^*||_p$$

$$= c_p ||P_m(\sum_{j=1}^{2} T_j^*YT_j^*) P_m||_p.$$

Since if a sequence $\{L_n\}$ of operators converges uniformly to L and $||L_n||_p \leq M$ then $||L||_p \leq M$, the proof is complete.

3. Recall that an operator A on a Banach space S is said to be Hermitian if $||e^{iaA}|| = 1$ for all real numbers. (Here the norm used is that of the Banach space S.) The operator A is said to be normal if A=P+iQ for some Hermitian operators P and Q such that [P,Q]=0. It is known [2,Theorem 2.2] that if A=P+iQ is a normal operator such that $As_n \to 0$ for a bounded seuence $\{s_n\} \in S$, then $Ps_n \to 0$ and $Qs_n \to 0$. Using this result it is easily seen that if A_1 and A_2 (:H -> H) are normal operators and $\{X_n\}$ is a bounded sequence in C_p, then $||A_1X_n+X_nA_2||_p \to 0$ implies $||A_1^*X_n+X_nA_2^*||_p \to 0$. It is clear from Theorem 1 that if A_1 and A_2 ($\in B(H)$) are commuting normal operators and $\{X_n\} \in C_p$, then

$$||\sum_{j=1}^{2} A_jX_nA_j||_p \to 0 \text{ implies } ||\sum_{j=1}^{2} A_j^*X_nA_j^*||_p \to 0 \text{ for } 2 \le p < \infty.$$ Does

this result hold for general commuting normal operators A_1 and A_2, and $\{X_n\} \in C_p$ for all p ? Although we can not at the moment answer this question we do have the following interesting result.

The operator A, on the Banach space S, is said to be hyponormal if A=P+iQ for some Hermitian operators P and Q such that the commutator $i(PQ-QP) \ge 0$ (i.e the numerical range of the commutator is contained in the non-negative reals).

THEOREM 2. Let A_j and B_j^*, j=1,2, be hyponormal operators such that $[A_1,A_2^*]=[B_1,B_2^*]=0$, and let $\{X_n\}$ be a bounded sequence in C_p.

If both $||\sum_{j=1}^{2} A_jX_nB_j||_p$ and $||A_1A_2X_n+X_nB_1B_2||_p$ converge to 0, then both $||\sum_{j=1}^{2} A_j^*X_nB_j^*||_p$ and $||A_2^*A_1^*X_n+X_nB_2^*B_1^*||_p$ converge to 0.

PROOF. Considering operators $T_j = A_j \oplus B_j^*$ and $Y_n = \begin{bmatrix} 0 & X_n \\ 0 & 0 \end{bmatrix}$ it is seen that it will suffice to prove the following: If T_1 and T_2 are hyponormal operators such that $[T_1, T_2^*]=0$, then $||T_1Y_nT_1^*+T_2Y_nT_2^*||_p$ and $||T_1T_2Y_n+Y_nT_1^*T_2^*||_p$ converge to 0 implies $||T_1^*Y_nT_1+T_2^*Y_nT_2||_p$ and $||T_2^*T_1^*X_n+X_nT_2T_1||_p$ converge to 0. Set

$$A=\begin{bmatrix} 0 & T_1 \\ T_1 & 0 \end{bmatrix}, B=\begin{bmatrix} 0 & T_2^* \\ T_2^* & 0 \end{bmatrix} \text{ and } W_n=\begin{bmatrix} 0 & T_2Y_n \\ Y_nT_1^* & 0 \end{bmatrix} \text{ on } \hat{H}=H\oplus H. \text{ Then}$$

A, B^* are hyponormal and $W_n \in C_p(\hat{H})$. define the operator $D_{A,B}: C_p(\hat{H}) \to C_p(\hat{H})$ by

$$\bar{D}_{A,B}W_n = AW_n + W_nB;$$

then, $D_{A,B}$ is hyponormal [3,Lemma 4.2] and $||D_{A,B}W_n||_{C_p(\hat{H})} = ||T_1Y_nT_1^*+T_2Y_nT_2^*||_p + ||T_1T_2Y_n+Y_nT_1^*T_2^*||_p$. Let $\overline{D}_{A,B}$ denote the (Banach space) adjoint of $D_{A,B}$. It is then clear that

$$\bar{D}_{A,B}W_n = A^*W_n + W_nB^*.$$

Applying [3, Corollary 4.6] we conclude that $||D_{A,B}W_n||_{C_p(\hat{H})} \to 0$.

Now define $V_n \in C_p(\hat{H})$ by $V_n = \begin{bmatrix} 0 & T_1^*Y_n \\ Y_nT_2 & 0 \end{bmatrix}$;

then, since $[T_1, T_2^*]=0$,

$$||\bar{D}_{A,B}W_n||_{C_p(\hat{H})} = ||D_{B^*,A^*}V_n||_{C_p(\hat{H})}.$$

The operator D_{B^*,A^*} (clearly) being hyponormal, another application of [3, Corollary 4.6] gives us that

$$\|\overline{D}_{B^*,A^*} V_n\|_{C_p(\hat{H})} = \|D_{B,A}V_n\|_{C_p(\hat{H})} = \|T_1^*Y_nT_1+T_2^*Y_nT_2\|_p + \|T_2^*T_1^*Y_n+Y_nT_2T_1\|_p$$

$$\to 0.$$

This completes the proof.

REFERENCES

1. K.Abdessemed and E.B.Davies, Some commutator estimates in the Schatten classes, J.Lond.Math.Soc. 39(1989), 297-308.

2. K.Mattila, Normal operators and proper boundary points of the spectra of operators on a Banach space, Ann.Acad.Sci.Fen. Ser.AI,Math.Dissertationes 19(1978).

3. K.Mattila, Complex strict and uniform convexity and hyponormal operators, Math. Proc.Camb.Phil.Soc. 96(1984), 483-493.

4. R.L.Moore and G.Weiss, The metric Fuglede property and normality, Canad.J.Math. 35(1983), 516-512.

5. V.S.Shul´man, On linear equation with normal coefficients, Sovt. Math. Dokl.27 (1983), 726-729.

6. G.Weiss, The Fuglede commutativity theorem modulo the Hilbert-Schmidt class and generating functions for matrix operators I, Trans.Amer.Math.Soc. 246(1978), 193-209; II, J.Operator Theory 5(1981), 3-16.

7. G.Weiss, The Fuglede commutativity theorem mudulo operator ideals, Proc.Amer.Math.Soc. 83(1981), 133-118.

8. G.Weiss, An extension of the Fuglede commutativity theorem modulo the Hilbert-Schmit class to operators of the form $\sum M_nXN_n$, Trans.Amer.Math.Soc. 278(1983), 1-20.

SCHOOL OF MATHEMATICAL SCIENCES
UNIVERSITY OF KHARTOUM
P.O.BOX 321
KHARTOUM
SUDAN.

DE BRANGES MODULES IN $H^2(\mathbb{C}^K)$ OF THE TORUS

B.S.YADAV. DINESH SINGH AND SANJEEV AGRAWAL

DEDICATED TO THE MEMORY OF U.N.SINGH

1. INTRODUCTION: In 1949, the fundamental paper of Beurling [1] appeared in which he characterized the invariant subspaces of the so called shift operator on the Hardy space H^2 of the unit circle. This characterization, which has become famous as Beurling's Theorem, sparked off a tremendous amount of research in numerous directions and it is now a corner stone of that area of harmonic and functional analysis where function theory and operator theory play major roles. We refer to [2],[3], [4], [5], [6],[9] and the works contained in numerous other books and journals. One of the lines of research has been to replace the Hardy space of scalar valued functions with vector valued functions, (see [5],[6] and [9]) while another approach has been to extend the characterization of Beurling to the situation of Hardy spaces of scalar valued functions on the torus, see [7],[10] and [13]. A third and recent approach, due to de-Branges [3], has further extended not only Beurling's original theorem but also its vector valued generalizations due to Lax and Halmos, for which we refer to [5] and [9].

In a recent paper, Singh [13] has extended the concept of de Branges spaces to the situation of scalar valued Hardy spaces of the torus. This characterization is quite similar to de Branges's theorem on the unit circle and includes as a special case the extension of Beurling's theorem to the torus (see[7] & [8]).

The purpose of this paper is to give necessary and sufficient conditions for a similar kind of characterization to hold in the situation of the $\mathbb{C}^k$- valued Hardy spaces on the torus. In other words we characterize those invariant subspaces of the $\mathbb{C}^k$- valued Hardy spaces of the torus that look like the subspaces in the theorems of de Branges and Beurling. The answer is neatly expressed in terms of $\mathbb{C}^k$ modules as we show below after putting down the necessary technical preliminaries.

2. PRELIMINARIES: Let A be a bounded linear transformation (operator) on a Hilbert space H. An invariant subspace of A is a proper non-trivial closed subspace M of H such that $A(M) \subset M$.

Let $H^2(T)$ be the set of all formal power series, $\sum_{n=0}^{\infty} \alpha_n z^n$, where the α_n are complex numbers such that $\sum_{n=0}^{\infty} |\alpha_n|^2 < \infty$. It is well known that $H^2(T)$ is a Hilbert space under the norm given by:

$$\text{If } f(z) = \sum_{n=0}^{\infty} \alpha_n z^n \in H^2(T), \text{ then } \|f(z)\|_2 = \left(\sum_{n=0}^{\infty} |\alpha_n|^2\right)^{1/2}$$

For f, $g \in H^2(T)$, f.g is defined as the Cauchy product of these two formal series. It is not necessary that $f.g \in H^2(T)$. $H^{\infty}(T)$ will denote the set of those f in $H^2(T)$ for which $f.g \in H^2(T)$ for all $g \in H^2(T)$.

DEFINITION 2.1. Let $\mathbb{C}^k$ denote the k-dimensional unitary space.

We now define:

(a) The product of $a=(a_1,\ldots\ldots,a_k)$ and $b=(b_1,\ldots\ldots,b_k)$ as $a.b = (a_1b_1,\ldots\ldots,a_kb_k)$

(b) $\|a\|_k^2 := |a_1|^2 + \ldots + |a_k|^2$

$H^2(\mathbb{C}^k)$ will denote the space of all square summable formal power series in z and w with coefficients from $\mathbb{C}^k$. That is, $f \in H^2(\mathbb{C}^k)$ is given by

$$f=f(z,w)=\sum_i \sum_j a_{ij} z^i w^j; \qquad \text{where } \sum_i \sum_j ||a_{ij}||_k^2 < \infty$$

[All summations in this paper are from 0 to ∞ unless otherwise stated].

We define

$$||f||_{H^2}=(\sum_i \sum_j ||a_{ij}||_k^2)^{1/2}$$

It is easy to check that $H^2(\mathbb{C}^k)$ is a Hilbert-space under the given norm. $H^\infty(\mathbb{C}^k)$ will, as before, denote the space of all of $f \in H^2(\mathbb{C}^k)$ such that $f.g \in H^2(\mathbb{C}^k)$ for all $g \in H^2(\mathbb{C}^k)$.We shall write H^2 for $H^2(\mathbb{C}^k)$ and H^∞ for $H^\infty(\mathbb{C}^k)$.

It is easy to see that H^2 is a $\mathbb{C}^k$-module under the multiplication:

$$a \in \mathbb{C}^k, \quad f=\sum_i \sum_j a_{ij} z^i w^j \in H^2$$
$$\Rightarrow af = \sum_i \sum_j aa_{ij} z^i w^j$$

It is useful to note that $H^2(T)$ and H^2 are in fact (isomorphic to) spaces of analytic functions on the open unit disk D and DxD respectively. See [6] and [10]. Also note that

$H^\infty(T)$ and H^∞ are Banach algebras under the norm

$$||f||_{H^\infty} = \sup\{ ||fg||_{H^2} : ||g||_{H^2} = 1\}$$

Let S denote the operator of multiplication by the coordinate function z on $H^2(T)$ i.e. $S(f(z))= zf(z)$. Further, let S and T denote the operators of multiplication by the coordinate functions z and w respectively on H^2, i.e.,

$$S(f(z,w)) = zf(z,w) \text{ and } T(f(z,w)) = wf(z,w)$$

It is easy to see that S on $H^2(T)$ and S and T on H^2 are isometries and that S and T commute on H^2. Two commuting operators A and B on a Hilbert Space H are said to be doubly commuting if $AB^*=B^*A$ Thus the commuting isometries S and T on H^2 are in fact doubly commuting on H^2.

DEFINITION 2.2. A Hilbert space M is said to be contractively contained in a Hilbert space H if

(i) M is a vector subspace of H

(ii) The inclusion map is a contraction, i.e.,

$$||x||_H \leq ||x||_M., \quad x \in M$$

BEURLING'S THEOREM [1]: Let M be an invariant subspace of S in $H^2(T)$. Then there exists an inner function q(z) in $H^\infty(T)$, (i.e.$\{q(z)z^n\}_{n=o}^{\infty}$ is an orthonormal set in $H^2(T)$) such that

$$M = q(z)H^2(T)$$

L.de Branges has given a generalisation of Beurling's Theorem in[3]. The scalar valued version (see[12]) is as follows:

DE BRANGE'S THEOREM: Let M be a Hilbert space contractively contained in $H^2(T)$ such that $S(M) \subset M$ and S acts as an isometry on M. Then there exists a b in the unit ball of $H^\infty(T)$ (unique upto a factor of unit modulus) such that

$$M = b(z)H^2(T)$$

and the norm in M is given by

$$||b(z)f(z)||_M = ||f||_2.$$

It is too much to expect an easy and nice characterisation of invariant subspaces of the operators S and T in the case of H^2. In fact, even in the scalar case (k=1), the common invariant subspaces of H^2 are not known. It is easy to produce examples which do not have the form $q(z,w)H^2$, i.e., similar to the form given by Beurling's Theorem, where $q(z,w)$ is an inner function (which means that $\{q(z,w)z^m w^n\}$ is an orthonormal system). Thus the case of general k will be that much more complicated and the same will be true if we seek to characterize de Branges - type spaces in H^2. In [13], Singh has given necessary and sufficient conditions for a space to be a de Branges type of space for k=1. Over here we shall extend this characterization to the situation of general k. In fact while looking at de Branges's characterization we shall be dropping the contractivity assumption. Actually there will be no continuity assumptions.

We recall that H^2 is a $\mathbb{C}^k$-module. We look at (algebraic) submodules of H^2.

DEFINITION 2.3. A closed subspace M of H^2 is said to be a Beurling Module if it is such that

$$M=bH^2, \quad b\in H^\infty$$

where $\{bz^m w^n\}$ is an orthonormal system.

REMARK 2.4. In case k=1, a Beurling module M turns out to be a Beurling space, i.e.,

$$M=qH^2(T^2)$$

where q is an inner function

DEFINITION 2.5. Let M be a vector subspace. Let M be a Hilbert space under some norm, then M shall be called a de Branges module if $M=bH^2$ where $b \in H^\infty$

and $\{\alpha_{mn} bz^m w^n\}$ is an orthogonal set in M for every sequence $\{\alpha_{mn}\}$ in $\mathbb{C}^k$.

REMARK 2.6 In case k=1, de Branges modules are in fact the de Branges spaces mentioned earlier, except that we have dispensed with the contractivity condition: $||f||_2 \leq ||f||_M$. In fact we have dispensed with any type of continuity relations between the two spaces. This enables us to characterize a larger class of invariant submodules.

We now fix some notation. $<.,.>_{H^2}, <.,.>_M$ and $<.,.>_k$ will denote the inner products on H^2, M (a Hilbert space which is a subspace of H^2) and $\mathbb{C}^k$ respectively. $||.||_{H^2}$, $||.||_M$, $||.||_k$ will denote the norms in H^2, M and $\mathbb{C}^k$ respectively. $\{e_i\}_{i=1}^k$ will denote the canonical linear basis for $\mathbb{C}^k$. If $f \in H^2$, then $f_{(i)}$ will denote $e_i f$. Note that in $f_{(i)}$ all coefficients have zero entries in all coordinates except may be in the ith coordinate. Also note that

(a) $f = f_{(1)} + f_{(2)} + \dots\dots + f_{(k)}$

(b) If $\alpha = (\alpha_1, \alpha_2, \dots, \alpha_k)$, then
$\alpha f_{(i)} = \alpha_i f_{(i)}$

We shall be using the following version of Halmos-Wold decomposition for two commuting isometries.

LEMMA 2.7 Let V_1, V_2, to be two doubly commuting isometries on a Hilbert space H which are shifts. Then

(i) The semigroup $\{V_1^m \; V_2^n\}_{m,n \geq 0}$ is of type ´s´ (see [14, P.225])

(ii) $L_1 \cap L_2$ is a wandering subspace for the semi group $\{V_1^m V_2^n\}_{m,n\geq 0}$

(i.e. $V_1^m V_2^n (L_1 \cap L_2) \perp V_1^k V_2^s (L_1 \cap L_2)$ if $(m,n) \neq (k,s)$ and

$H = \sum \sum \oplus V_1^m V_2^n (L_1 \cap L_2)$ where $L_k = H \ominus V_k H$ $(k=1,2)$.

PROOF. See [14, Theorem 1]

LEMMA 2.8 :Let V_1, V_2 be commuting isometries on the Hilbert space H. If $L_1 \cap L_2 = \{0\}$ where $L_k = H \ominus V_k H$, then the semigroup $\{V_1^m V_2^n\}_{m,n\geq 0}$ is not of the type ´s´

PROOF. See [14, Corollary 1]

THE MAIN RESULT

THEOREM Let M be a Hilbert space such that M is a vector subspace H^2. Let S and T leave M invariant and act as isometries on it. Then M is a de Branges module if and only if S and T commute doubly on M and $((M \ominus S(M)) \cap (M \ominus T(M)))$ is a $\mathbb{C}^k$- submodule of H^2.

Let $N = (M \ominus S(M)) \cap (M \ominus T(M))$.

We first assume that S and T commute doubly on M and N is a $\mathbb{C}^k$-submodule of H^2.By Lemmas 2.7 and 2.8, $N \neq \{0\}$.We prove a few lemmas before providing the proof of the theorem.

LEMMA 3.1 $M = \sum_i \sum_j \oplus S^i T^j (N)$

PROOF This is a direct application of Lemma 2.7

LEMMA 3.2 If $g \in N$ and $||g||_M = 1$, then

(i) $\{g z^m w^n)$ is an orthonormal system and

(ii) $\{a_{mn} g z^m w^n\}$ is an orthogonal system for every sequence $\{a_{mn}\}$ in $\mathbb{C}^k$.

PROOF (i) The fact that S and T are isometries on M and Lemma 3.1 immediately prove this result

(ii) As N is a $\mathbb{C}^k$- submodule of H^2, $\alpha_{mn} g \in N$. Now Lemma 3.1 immediatel proves this result.

NOTATION. $\mathbb{C}^k f = \{af \mid a \in \mathbb{C}^k\} = \langle f \rangle$.

LEMMA 3.3 Let $f \in N$, then there exist positive constants A_f and B_f such that

$$A_f \,||\alpha f||_{H^2} \leq ||\alpha f||_M \leq B_f \,||\alpha f||_{H^2} \text{ for all } \alpha \in \mathbb{C}^k.$$

PROOF. Let $\alpha = (\alpha_1, \alpha_2, \ldots, \alpha_k)$, $\alpha_i \in \mathbb{C}$.

Then, $\alpha f = \alpha_1 f_{(1)} + \ldots + \alpha_k f_{(k)}$.

As $f_{(i)} = e_i f \in \langle f \rangle$, we see that $\langle f \rangle$ is a finite dimensional vector subspace of M and of H^2. Thus it is a closed subspace of M and of H^2. As it is a finite dimensional space, all norms on it are equivalent and hence there exist positive constants A_f and B_f such that

$$A_f \,||\alpha f||_{H^2} \leq ||\alpha f||_M \leq B_f \,||\alpha f||_{H^2} \text{ for all } \alpha \in \mathbb{C}^k.$$

LEMMA 3.4 $f \in N \Rightarrow f.g \in M$, for all $g \in H^2$

PROOF Let $f(z,w) = \sum_i \sum_j a_{ij} z^i w^j$

$$g(z,w) = \sum_i \sum_j b_{ij} \; z^i w^j$$

$$Y_t = \sum_{i=0}^{t} b_{i,t-i} \; z^i w^{t-i}$$

$$P_n = Y_o + Y_1 + \ldots + Y_n$$

It is easy to see that $P_n \to g$ in H^2. Hence $\{P_n\}$ is a Cauchy sequence in H^2. Thus given $\epsilon > 0$, there exists $N > 0$ such that

$$||P_m - P_n||^2_{H^2} < \epsilon^2, \text{ for all } m,n > N$$

Now $\{b_{ij}z^i w^j\}$ is an orthogonal set in H^2 and $||b_{ij}z^i w^j||^2_{H^2} = ||b_{ij}||^2_k$.

Thus, for all $m > n > N$ we have

$$\epsilon^2 > ||P_m - P_n||^2_{H^2} = ||Y_{n+1} + \ldots + Y_m||^2_{H^2}$$

$$= ||Y_{n+1}||^2_{H^2} + \ldots + ||Y_m||^2_{H^2}$$

$$= \sum_{i=0}^{n+1} ||b_{i,n+1-i}||^2_k + \ldots + \sum_{i=o}^{m} ||b_{i,m-i}||^2_k \qquad (1)$$

Thus

$$||P_m f - P_n f||^2_M = ||\sum_{i=o}^{n+1} b_{i,n+1-i} z^i w^{n+1-i} f + \ldots + \sum_{i=o}^{m} b_{i,m-i} z^i w^{m-i} f||^2_M$$

$$= \sum_{i=o}^{n+1} ||b_{i,n+1-i} z^i w^{n+1-i} f||^2_M + \ldots + \sum_{i=o}^{m} ||b_{i,m-i} z^i w^{m-i} f||^2_M$$

(by Lemma 3.2)

$$= \sum_{i=o}^{n+1} ||b_{i,n+1-i} f||^2_M + \ldots + \sum_{i=o}^{m} ||b_{i,m-i} f||^2_M$$

(as S and T are isometries on M)

$$\le (B_f)^2 \Big(\sum_{i=0}^{n+1} ||b_{i,n+1-i} f||^2_{H^2} + \ldots + \sum_{i=0}^{m} ||b_{i,m-i} f||^2_{H^2} \Big)$$

(by Lemma 3.3)

$$\le (B_f\|f\|_{H^2})^2 \left(\sum_{i=0}^{n+1} \|b_{i,n+1-i}\|_k^2 + \ldots + \sum_{i=0}^{m} \|b_{i,m-i}\|_k^2\right)$$

$$(\text{As } \|\alpha f\|_{H^2} \le \|\alpha\|_k \|f\|_{H^2})$$

$$\le (B_f\|f\|_{H^2})^2 \in^2$$

(by (1))

Thus $\{P_n f\}$ is Cauchy in M and hence converges, to say h, in M,

As $h \in H^2$, let it be

$$h = \sum_i \sum_j h_{ij} z^i w^j$$

We show $h_{ij} = \sum_{r=o} \sum_{s=o} b_{rs}\, a_{i-r,j-s}$ (2)

This shall imply that h = fg.

Fix i and j. Let i+j = t and n > t.

Then $P_n = Y_o + Y_1 + \ldots + Y_t + Y_{t+1} + \ldots + Y_n$

$$\text{Now } Y_{t+1}f + \ldots + Y_n f = \sum_{r=t+1}^{n} \sum_{i=o}^{r} b_{i,r-i} z^i w^{r-i} f$$

$$= w^{t+1}(b_{o,t+1} + b_{o,t+2}w + \ldots + b_{o,n}w^{n-t-1})f$$

$$+ zw^t(b_{1,t} + b_{1,t+1}w + \ldots + b_{1,n-1}w^{n-t-1})f$$

+
:
:
:
+

$$+ z^t w(b_{t,1} + b_{t,2}w + \ldots + b_{t,n-t}w^{n-t-1})f$$

$$+ z^{t+1}(b_{t+1,0} + \sum_{i=t+1}^{t+2} b_{i,t+2-i} z^{i-t-1} w^{t+2-i} + \ldots$$

$$\ldots + \sum_{i=t+1}^{n} b_{i,n-i} z^{i-t-1} w^{n-i})f$$

$$= w^{t+1} Q_{0,n} + zw^{t} Q_{1,n} + \ldots + z^{t+1} Q_{t+1,n}$$

But $P_n f$ converges to h in M and hence $P_n f - P_t f$ converges to $h - P_t f$ in M.

Thus $\{w^{t+1} Q_{0,n} + zw^{t} Q_{1,n} + \ldots + z^{t+1} Q_{t+1,n}\}$ is Cauchy in M.

But S and T are isometries on M and $\{\alpha_{mn} z^m w^n f\}$ is an orthogonal set. This easily shows that, for each r = 0,1,2,...,t+1;$\{Q_{r,n}\}$is Cauchy in M. Let them converge to $\phi_0, \phi_1, \phi_2, \ldots, \phi_{t+1}$ respectively. Then, again, as S and T are isometries on M, we have

$$z^r w^{t+1-r} Q_{r,n} \text{ converges to } z^r w^{t+1-r} \phi_r$$

Thus $h - P_t f = w^{t+1} \phi_0 + zw^{t} \phi_1 + \ldots + z^{t+1} \phi_{t+1}$

But there is no term with $z^i w^j$ on the RHS (as for each term on the RHS, the indices add up to greater than or equal to $t+1 > i+j$).

Thus the $(i,j)^{th}$ term of h

$= (i,j)^{th}$ term of $P_t f$

i.e. $h_{ij} = \sum_{r=o}^{i} \sum_{s=o}^{j} b_{rs}\, a_{i-r,j-s}$

This is what we required (see(2))

Note that we have shown that if $g \in N$ and $f = \sum_i \sum_j a_{ij} z^i w^j \in H^2$, then

$$(\sum_i \sum_j a_{ij} z^i w^j g) = (\sum_i \sum_j a_{ij} z^i w^j) g$$

$$= f.g$$

LEMMA 3.5. If $f, g \in N$ are such that $f_{(i)} \perp g_{(i)}$, then either $f_{(i)}=0$ or $g_{(i)}=0$.

PROOF. We first show that $f_{(i)}H^2 \perp g_{(i)}H^2$.

Let $h = \sum_m \sum_n a_{mn} z^m w^n$ and $k = \sum_m \sum_n b_{mn} z^m w^n$ belong to H^2 where $a_{mn} = (a_{mn1}, \ldots, a_{mnk})$ and $b_{mn} = (b_{mn1}, \ldots, b_{mnk})$.

Note that $f_{(i)}h = f_{(i)}h_{(i)}$ and $g_{(i)}k = g_{(i)}k_{(i)}$

Also $h_{(i)} = \sum_m \sum_n a_{mni} z^m w^n$ and $k_{(i)} = \sum_m \sum_n b_{mni} z^m w^n$.

As S and T are isometries on M; $a_{mni}f_{(i)}$ and $b_{mni}g_{(i)} \in N$; and $f_{(i)} \perp g_{(i)}$ we have

$< a_{mni} z^m w^n f_{(i)}, b_{rsi} z^r w^s g_{(i)} >_M = 0$

Thus, it follows quite easily that

$< hf_{(i)}, kg_{(i)} >_M = 0.$

So, $f_{(i)}H^2 \perp g_{(i)}H^2$.

But $f_{(i)}g_{(i)} \in f_{(i)}H^2 \cap g_{(i)}H^2$

Thus $f_{(i)}g_{(i)}=0$

We now "define" the "first" non zero coefficient of $f \in H$.

Let $$f = \sum_i \sum_j a_{ij} z^i w^j$$

Then a_{rs} will be called the first non zero coefficient of f if and only if

(i) $a_{rs} \neq 0$

(ii) $a_{ij} = 0$ for all i,j such that $i+j < r+s$

(iii) $a_{ij} = 0$ for all i,j such that $i+j = r+s$ and $i < r$

It is obvious that for non zero f, the first non zero coefficient is uniquely determined. In fact, if,

$$Y_n = \sum_{i+j=n} a_{ij} z^i w^j$$

then the first non zero coefficient of f occurs in the first (least n) non zero Y_n. It is also the coefficient of the non zero term in Y_n with least value of the first subscript.

Let a_{rs} and b_{tm} be the first non zero coefficient of f and g where

$$f = \sum_n f_n \quad , \quad f_n = \sum_{i+j=n} a_{ij} z^i w^j$$

$$g = \sum_n g_n \quad , \quad g_n = \sum_{i+j=n} b_{ij} z^i w^j$$

Let $f.g = \sum_i \sum_j c_{ij} z^i w^j$

We show that

(i) $c_{ij} = 0$ for all $i+j < r+t+s+m$

(ii) $c_{r+t,s+m} = a_{rs} b_{tm}$

Now $f.g = \sum_n X_n$

where $X_n = f_o g_n + \ldots + f_n g_o$ Note that $c_{ij} z^i w^j$ occurs in X_{i+j}

As the first non zero coefficients of f and g are a_{rs} and b_{tm}

respectively, we have

$$f_o = f_1 = \dots = f_{r+s-1} = 0$$

$$g_o = g_1 = \dots = g_{t+m-1} = 0 \qquad (3)$$

Now consider $f_i g_j$, where $i+j < r+t+s+m$ Then either $i < r+s$ or $j < t+m$

Thus either $f_i = 0$ or $g_j = 0$

which means $f_i g_j = 0$ for all $i+j < r+t+s+m$

But $X_k = \sum_{i+j=k} f_i \, g_j$

So $X_k = 0$ for all $k < r+t+s+m$ (4)

Let $N = r+t+s+m$

$$X_N = f_N g_o + f_{N-1} g_1 + \dots + f_{r+s+1} g_{t+m-1} + f_{r+s} g_{t+m}$$
$$+ f_{r+s-1} g_{t+m+1} + \dots + f_o g_N$$
$$= f_{r+s} g_{t+m} \qquad \text{(by(3))}$$

$$= \sum \sum a_{ij} b_{pq} \; z^{i+p} \; w^{j+q}$$
(The first summation is over $i+j = r+s$ and the second is over $p+q = t+m$)

Thus the coefficient of $z^{r+t} w^{s+m}$ is

$$\sum a_{ij} \; b_{pq} \qquad (5)$$

where the sum is taken over i,j,p,q such that

$$\begin{aligned} i+j &= r+s \\ p+q &= t+m \\ i+p &= r+t \\ j+q &= s+m \end{aligned} \qquad (6)$$

Now if $i < r$, then $a_{ij} = 0$ if $i+j = r+s$

and if $p < t$, then $b_{pq} = 0$ if $p+q = t+m$

(as a_{rs} and b_{tm} are the first non zero coefficients of f and g respectively).

Thus $a_{ij} b_{pq} = 0$ if $i < r$ or $p < t$ and $i+j = r+s$ and $p+q = t+m$

So, in (5), the only possible non zero entries are where $i \geq r$

and $p \geq t$.

But $i+p = r+t$. This means $i=r$ and $p=t$.
Thus the only non zero summand in (5) could be $a_{rs}b_{tm}$.

$$\text{So } c_{r+t,s+m} = a_{rs}b_{tm}$$

Now, let if possible, $f_{(i)} \neq 0$ and $g_{(i)} \neq 0$. Let a_{rs} and b_{tm} be the first non-zero coefficients of $f_{(i)}$ and $g_{(i)}$ respectively.

Then $a_{rsi} \neq 0$ and $b_{tmi} \neq 0$. Also $a_{rs}b_{tm}$ is the coefficient of $z^{r+t}w^{s+m}$ in $f_{(i)}g_{(i)}$.

Thus, $a_{rs}b_{tm} = 0$
Hence, $a_{rsi}b_{tmi} = 0$
This gives us a contradiction.
Hence either $f_{(i)} = 0$ or $g_{(i)} = 0$.

PROOF OF THEOREM.

Let i be such that there exists $f \in N$ where $f_{(i)} \neq 0$. Let $g \in N$. We show that $g_{(i)}$ is linearly dependent on $f_{(i)}$.

$$\text{Let } h_{(i)} = g_{(i)} - \frac{\langle g_{(i)}, f_{(i)} \rangle_M}{\| f_{(i)} \|_M^2} f_{(i)}.$$

Then $h_{(i)} \in N$ and $h_{(i)} \perp f_{(i)}$. As $f_{(i)} \neq 0$, we have by Lemma 3.5 that $h_{(i)}=0$

$$\text{or} \qquad g_{(i)} = \frac{\langle g_{(i)}, f_{(i)} \rangle_M}{\| f_{(i)} \|_M^2} f_{(i)}.$$

Let $X = \{i \mid \text{There exists } f \in N \text{ such that } f_{(i)} \neq 0\}$.

For each i in X, choose $b_{(i)} \in N$ such that $b_{(i)} \neq 0$.

Let $b = \sum_{i \in X} b_{(i)}$.

We show $N = \langle b\rangle$.

Let $g \in N$. Then $g = \sum_{i\in X} g_{(i)}$.

But there exist $\alpha_i \in \mathbb{C}$ such that $g_{(i)} = \alpha_i b_{(i)}$.

Let $\beta = (\beta_1, \beta_2, \ldots, \beta_k)$ where

$$\beta_i = \begin{cases} \alpha_i & , \ i \in X \\ 0 & , \ i \notin X. \end{cases}$$

Then $g = \beta b$ and hence $N = \langle b\rangle$.

We show that $M = bH^2$.

By Lemma 3.3, $bH^2 \subset M$.

Let $f \in M$.

By Lemma 3.1, there exist $\alpha_{mn} \in \mathbb{C}^k$ such that

$$f = \sum_m \sum_n \alpha_{mn} z^m w^n b$$

Now, for $i \notin X$, $b_{(i)} = 0$, thus $f_{(i)} = 0$. So, without loss of generality we can assume that $X = \{1,2,\ldots,k\}$, i.e., $b_{(i)} \neq 0$ for each i.

We show that $\sum_m \sum_n \alpha_{mn} z^m w^n \in H^2$.

$$\infty > \|f\|_M^2 = \|\sum_m \sum_n \alpha_{mn} z^m w^n b\|_M^2$$

$$= \sum_m \sum_n \|\alpha_{mn} z^m w^n b\|_M^2$$

(by Lemma 3.2)

$$= \sum_m \sum_n \|\alpha_{mn} b\|_M^2$$

(As S and T are isometries on M)

$$\geq A_b^2 \sum_m \sum_n \|\alpha_{mn} b\|_{H^2}^2$$

(Lemma 3.3)

$$= A_b^2 \sum_m \sum_n \left(\sum_{i=1}^k |\alpha_{mni}|^2 \|b_{(i)}\|_{H^2}^2 \right)$$

$$= A_b^2 \sum_{i=1}^{k} \left(\sum_m \sum_n |\alpha_{mni}|^2 \right) ||b_{(i)}||_{H^2}^2$$

Thus, for each i, $\sum_m \sum_n |\alpha_{mni}|^2 < \infty$.

So, $$\sum_{i=1}^{k} \sum_m \sum_n |\alpha_{mni}|^2 < \infty$$

or, $$\sum_m \sum_n ||\alpha_{mn}||_k^2 < \infty$$

Thus, if $g = \sum \sum \alpha_{mn} z^m w^n$, then $g \in H^2$.

By Lemma 3.4, $bg = \sum_m \sum_n \alpha_{mn} z^m w^n b$

$= f.$

Thus $f \in bH^2$ or $M \subset bH^2$.

Note that $\{ a_{mn} z^m w^n b\}$ is an orthogonal set in M for every sequence $\{a_{mn}\}$ in M.

Thus M is a de Branges module.

For the converse, assume that M is a vector subspace of H^2 and let M be invariant under S and T which act as isometries on it. Let $M = bH^2$ where $\{a_{mn} bz^m w^n\}$ is an orthogonal set in M for every sequence $\{a_{mn}\}$ in $\mathbb{C}^k$. It is obvious that M is a $\mathbb{C}^k$- submodule of H^2. We show that S and T commute doubly on M. S and T commute on M. We now determine T^* (on M). Let $\alpha, \beta \in \mathbb{C}^k$.

$$\langle T^* \alpha z^m w^n b, \beta z^r w^s b \rangle_M$$

$$= \langle \alpha z^m w^n b, T(\beta z^r w^s b) \rangle_M$$

$$= \langle \alpha z^m w^n b, \beta z^r w^{s+1} b \rangle_M$$

$$= \begin{cases} 0 & \text{if } n \neq s+1 \text{ or } m \neq r \\ \langle \alpha b, \beta b \rangle_M & \text{otherwise} \end{cases}$$

[as $\{\alpha_{mn} z^m w^n\}$ is an orthogonal set in M and S and T act as isometries on it].

$$= \begin{cases} \langle \alpha z^m w^{n-1} b \, , \beta z^r w^s b \rangle_M & , \quad n \geq 1 \\ \langle 0 \quad , \quad \beta z^r w^s b \rangle_M & , \quad n = 0 \end{cases}$$

$$\text{Thus } T^*(\alpha b z^m w^n) = \begin{cases} \alpha z^m w^{n-1} b & n \geq 1 \\ 0 & n = 0 \end{cases}$$

$$\text{Thus } ST^*(\alpha b z^m w^n) = \begin{cases} \alpha z^{m+1} w^{n-1} b & n \geq 1 \\ 0 & n = 0 \end{cases}$$

$$= T^*(\alpha z^{m+1} w^n b)$$

$$= T^* S(\alpha z^m w^n b)$$

So S and T commute doubly on the set $\{\alpha_{mn} z^m w^n b\}$ which generates M (as $M = bH^2$), [by a proof similar to that of Lemma 3.4].

Hence S and T commute doubly on M.

We now show that $N = \langle b \rangle$ and hence N is a $\mathbb{C}^k$- submodule of H^2.

We first show that $\alpha b \in N$ for every $\alpha \in \mathbb{C}^k$.

Let $f = \sum_m \sum_n \alpha_{mn} z^m w^n \in H^2$,

and $Y_t = \sum_{i=0}^{t} \alpha_{i,t-i} z^i w^{t-i}$

and $f_n = Y_0 + Y_1 + \dots + Y_n$

Then, as in Lemma 3.4, $bf_n \to bf$ in M.

As S is an isometry on M,

$$zbf_n \to zbf \text{ in } M.$$

Thus $\langle \alpha b, zbf_n \rangle_M \to \langle \alpha b, zbf \rangle_M$

But $\langle \alpha b, zbf_n \rangle_M = \sum_{i=0}^{n} \langle \alpha b, zbY_i \rangle_M$

$$= \sum_{i=0}^{n} \sum_{j=0}^{i} \langle \alpha b, \alpha_{j,i-j} z^{j+1} w^{i-j} b \rangle_M$$

$$= 0$$

[As αb is orthogonal to the set $\{\alpha_{mn} z^m w^n\}_{m\geq 1}$].

Hence, $\langle \alpha b, zbf \rangle_M = 0$.

Thus, $\alpha b \in M \ominus S(M)$.

Similarly, $\alpha b \in M \ominus T(M)$.

Thus, $\alpha b \in N$ or $\langle b \rangle \subset N$.

If $g \in N \ominus \langle b \rangle$, let $g = bh$ where $b \in H^2$ and

$$h = \sum_m \sum_n h_{mn} z^m w^n; \qquad Y_t = \sum_{i=0}^{t} h_{i,t-i} z^i w^{t-i} \quad \text{and} \quad h_n = Y_0 + Y_1 + \ldots + Y_n.$$

As above, $bh_n \to bh = g$ in M.

Now, $\langle h_{oo}b, bh_n \rangle_M = \langle h_{oo}b, h_{oo}b \rangle_M = ||h_{oo}b||_M^2$.

But $\langle h_{oo}b, bh_n \rangle_M \to \langle h_{oo}b, bh \rangle_M$.

Thus $h_{oo}b = 0$

So $g = b(zh_1 + wh_2)$ where

$$h_1 = \sum_{i=1}^{\infty} h_{io} z^{i-1} \quad \text{and} \quad h_2 = \sum_{i=0}^{\infty} \sum_{j=1}^{\infty} h_{ij} z^i w^{j-1}$$

But as $g \in N$, $\langle g, bzh_1 \rangle_M = 0$ and $\langle g, bwh_2 \rangle_M = 0$.

Thus $g = 0$.

Hence $N = \langle b \rangle$.

REFERENCES

1) A. Beurling, On two problems concerning linear transformation in Hilbert space, Acta Math. 81(1949) 239-255.

2) L. de Branges and J. Rovnyak, Square Summable Power Series, Holt, Rhinehart and Wingston, 1967.

3) L. de Branges, Square Summable Power Series, Springer Verlag, (To appear).

4) J. Garnett, Bounded Analytic Functions, Academic Press, 1981.

5) H. Helson, Lectures on Invariant Subspaces, Academic Press, 1964.

6) K. Hoffman, Banach Spaces of Analytic Functions, Prentice Hall, 1962.

7) V. Mandrekar, The validity of Beurling theorems in polydiscs, Proc. Amer. math. Soc. 103 (1988) 145-148.

8) J. Radlow, Closed ideals in square summable power series. Proc. Amer. Math. Soc. 38(1973) 293-297.

9) M.Rosenblum and J. Rovnyak, Hardy Classes and Operator theory. Oxford University Press, 1985.

10) W. Rudin, Function Theory in Polydiscs, Benjamin, 1969.

11) D. Sarason, Function theory on the unit circle, Virginia Polytech. Inst. and State Univ. Blacksburg, VA, 1979.

12) D. Sarason, Shift invariant spaces from the Brangesian view point. (Proc. Sympos. on the occasion of the proof of the Bieberbach conjective, 1986) Math. Monogr. Amer. Math. Soc., Providence, RI 1986, pp 153-166.

13) D. Singh, Brangesian spaces in the polydisck, Proc. Amer. Math. Soc. 110 (1990) 971-977.

14) M. Slocinskii, On the Wold type decomposition of a pair of commuting isometries. Ann. Polon. math. 37(1980), 255-262.

B.S. Yadav and Dinesh Singh
Department of Mathematics,
University of Delhi,
Delhi - 110007,
India.

Sanjeev Agrawal
Department of Mathematics,
St. Stephen's College,
University of Delhi,
Delhi - 110007,
India.

Weak Compactness of Holomorphic Composition Operators on H^1

DONALD SARASON

In memory of U. N. Singh.

In [5] it was proved that the weak compactness of a holomorphic composition operator on L^1 of the unit circle implies its compactness, and the question was raised whether the analogous result holds in the Hardy space H^1. In this note the question will be answered in the affirmative. Although the ideas invol ed do not go substantially beyond those of [5], the result seems worth presenting, partly because, in conjunction with a recent result of J. H. Shapiro and C. Sundberg [6] and another result from [5], it has a rather curious corollary.

Let b be a holomorphic self-map of the unit disk, D, that vanishes at the origin but is not identically 0. The assumption that $b(0) = 0$ is a convenience which results in no loss of generality for the question addressed here. (This is explained in [6].) The composition operator C_b acts on a holomorphic function h in D in the obvious way: $(C_b h)(z) = h(b(z))$. It is well known that C_b acts as a bounded operator on each of the Hardy spaces H^p $(0 < p \leq \infty)$ [4].

It was pointed out in [5] that C_b can also be defined on the L^p spaces of normalized Lebesgue measure on ∂D $(1 \leq p \leq \infty)$. If h is a function in one of those spaces, we regard h as extended into D via Poisson's formula and define $C_b H$ by $(C_b h)(e^{i\theta}) = h(b(e^{i\theta}))$. Then, as is shown in [5], C_b acts as a bounded operator on each L^p, and the Poisson integral of $C_b h$ is the composition of b with the Poisson integral of h.

It will be convenient to work, not with the classical Hardy space H^1, but with its real-variable version, which will be denoted by H_1. The latter space consists of those functions h in L^1 such that $\tilde{h}$, the conjugate function of h, is also in L^1; its norm is defined by

$$\|h\|_{H_1} = \|h\|_1 + \|\tilde{h}\|_1.$$

One easily verifies that if h is in H_1 then $C_b \tilde{h}$ is the conjugate function of $C_b h$. Hence C_b acts as a bounded operator on H_1. Clearly, C_b is compact or weakly compact on H_1 if and only if it has the same property on H^1.

THEOREM. *If C_b is weakly compact on H_1 then C_b is compact on H_1.*

For the proof, some properties of the adjoint operator, C_b^*, are needed. Normalized Lebesgue measure on ∂D will be denoted by m. For h in L^p and g in $L^{p'}$, where $p' = p/(p-1)$, we define $\langle h, g\rangle = \int h\bar{g}\,dm$, and we recall that $L^{p'}$ is the dual of L^p under this pairing ($1 \leq p < \infty$). If $q < p$ then $L^{q'} \subset L^{p'}$, and the restriction to $L^{q'}$ of the adjoint of C_b as an operator on L^p coincides with the adjoint of C_b as an operator on L^q. We may thus speak without ambiguity of a single adjoint operator C_b^*, which acts on $\cup_{p>1} L^p$. The operator C_b^* maps trigonometric polynomials to trigonometric polynomials, and it maps C, the space of continuous complex-valued functions on ∂D, into itself [5].

The well-known duality theorem of C. Fefferman says that BMO, the space of functions of bounded mean oscillation on ∂D, is the dual of H_1 under the pairing above (except that, for a given function g in BMO, the product $h\bar{g}$ may not be integrable for all h in H_1 but only for a dense subset of H_1). Similarly, H_1 is the dual, under the same pairing, of VMO, the subspace of BMO of functions of vanishing mean oscillation. These matters are explained in Sections I.9 and I.11 of [2].

Because C_b commutes with the conjugation operator, one easily verifies hat C_b^* also commutes with the conjugation operator. Because BMO consists of the functions on ∂D writable as sums of functions in L^∞ and their conjugates, it follows that C_b^* maps BMO into itself. Similarly, because VMO consists of the functions on ∂D writable as sums of functions in C and their conjugates, it follows that C_b^* maps VMO into itself. One easily checks that C_b as an operator on H_1 is the adjoint of C_b^* as an operator on VMO, and C_b^* as an operator on BMO is the adjoint of C_b as an operator on H_1.

Two facts about weakly compact operators are needed. The first states that an operator T on a Banach space X is weakly compact if and only if T^{**} maps X^{**} into X ([1], p. 482). The second states that an operator is weakly compact if and only if its adjoint is ([1], p. 485). From these and the remarks above it follows that C_b is weakly compact on H_1 if and only if C_b^* maps BMO into VMO.

The first step in the proof of the theorem will be to show that if C_b is weakly compact on H_1 then $|b| < 1$ almost everywhere on ∂D. Once that has been accomplished, the rest will follow fairly quickly.

Let the Borel measure ν on ∂D be defined by $\nu(E) = m(b^{-1}(E))$. Showing that $|b| < 1$ almost everywhere on ∂D amounts to showing that ν is the zero measure. First it will be shown that ν is absolutely continuous with respect to m. (Although this fact is well known, the details are provided for completeness.) If ν were not absolutely continuous, there would be a Borel subset E of ∂D such that $m(E) = 0$ and $m(b^{-1}(E)) > 0$. Then there would by Lusin's theorem be a closed subset F of $b^{-1}(E)$ such that $m(F) > 0$ and such that $b|F$ is continuous. The set $b(F)$ would then be closed and of Lebesgue measure 0, so, by a well-known theorem of P. Fatou ([3], p. 80), there would be a nonzero function h in the disk algebra ($= H^\infty \cap C$) that vanishes on $b(F)$. Then the function $C_b h$ would be a nonzero function in H^∞ having the radial limit 0 at each point of the set F of positive Lebesgue measure, in violation of the boundary unicity theorem for H^∞. Hence ν is absolutely continuous with respect to m.

Now suppose that C_b is weakly compact on H_1 but that ν is not the zero measure. Then there is a Borel subset G of ∂D with $0 < m(G) < 1$ on which $\frac{d\nu}{dm}$ is bounded away from 0, say $\frac{d\nu}{dm} \geq \varepsilon > 0$ on G. Let $F = b^{-1}(G)$. It is asserted that $C_b^* \chi_F = \frac{d\nu}{dm}\chi_G$. To verify this, let E be any Borel subset of ∂D. Then

$$\begin{aligned}\langle \chi_E, C_b^* \chi_F\rangle = \langle C_b \chi_E, \chi_F\rangle &= \int_F \chi_E \circ b\, dm \\ &= m(b^{-1}(E \cap G)) = \nu(E \cap G) \\ &= \int_E \frac{d\nu}{dm}\chi_G dm = \langle \chi_E, \frac{d\nu}{dm}\chi_G\rangle,\end{aligned}$$

and the assertion follows.

As noted above, the weak compactness of C_b on H_1 implies that C_b^* maps BMO into VMO. Thus, to obtain a contradiction, it only remains to show that the function $\frac{d\nu}{dm}\chi_G$ is not in VMO. That will be accomplished, in standard fashion, by showing that there are arbitrarily short subarcs of ∂D on which the mean oscillation of that function is at least $\varepsilon/2$.

For λ in ∂D and $\delta > 0$, let $I(\lambda, \delta)$ denote the subarc of ∂D with center λ and normalized Lebesgue measure δ. For fixed δ, the function

$$\frac{1}{\delta}\int_{I(\lambda,\delta)} \frac{d\nu}{dm}\chi_G dm,$$

the average of the function $\frac{d\nu}{dm}\chi_G$ over $I(\lambda, \delta)$, is a continuous function of δ. If δ is small, this function takes values both smaller than $\varepsilon/2$ and larger than $\varepsilon/2$ (by the Lebesgue differentiation theorem). For such δ, then, there is a λ with

$$\frac{1}{\delta}\int_{I(\lambda,\delta)} \frac{d\nu}{dm}\chi_G dm = \varepsilon/2.$$

For such a λ we have, for the mean oscillation of $\frac{d\nu}{dm}\chi_G$ over $I(\lambda, \delta)$, the estimate

$$\begin{aligned}&\frac{1}{\delta}\int_{I(\lambda,\delta)} \left| \frac{d\nu}{dm}\chi_G - \frac{1}{\delta}\int_{I(\lambda,\delta)} \frac{d\nu}{dm}\chi_G dm \right| dm \\ &= \frac{1}{\delta}\int_{I(\lambda,\delta)} \left| \frac{d\nu}{dm}\chi_G - \frac{\varepsilon}{2}\right| dm \\ &\geq \frac{1}{\delta}\int_{I(\lambda,\delta)} \frac{\varepsilon}{2} dm = \frac{\varepsilon}{2},\end{aligned}$$

showing that $\frac{d\nu}{dm}\chi_G$ is not in VMO, the desired contradiction. We can conclude that ν is the zero measure and hence that $|b| < 1$ almost everywhere on ∂D.

To complete the proof of the theorem, let $(h_n)_1^\infty$ be a bounded sequence in H_1. We need to show, assuming C_b is weakly compact on H_1, that the sequence $(C_b h_n)_1^\infty$ has a subsequence that converges in the norm of H_1. Since H_1 is the dual of the separable space VMO, we can assume without loss of generality that the sequence $(h_n)_1^\infty$ converges in the weak-star topology of H_1, and that its weak-star limit is 0. Then $h_n \to 0$ pointwise in D, so $C_b h_n \to 0$ almost everywhere (because $|b| < 1$ almost everywhere on ∂D). By the weak compactness of C_b on H_1 there is a subsequence $(C_b h_{n_k})_{k=1}^\infty$ that converges to 0 in the weak topology of H_1 and hence also in the weak topology of L^1. Because a sequence in L^1 that converges both weakly and almost everywhere converges also in L^1 norm ([1], p. 295), we can conclude that $\|C_b h_{n_k}\|_1 \to 0$. Also, because the conjugation operator is bounded and hence weakly continuous on H_1, we have that $(C_b h_{n_k})^\sim \to 0$ weakly. And, because $\tilde{h}_n \to 0$ pointwise in D and $(C_b h_n)^\sim = C_b \tilde{h}_n$, we have that $(C_b h_n)^\sim \to 0$ almost everywhere on ∂D. Thus, we can also conclude that $\|(C_b h_{n_k})^\sim\|_1 \to 0$. Therefore $C_b h_{n_k} \to 0$ in H_1 norm, and the proof of the theorem is complete.

To state the corollary alluded to at the beginning of this note we introduce the space $\mathrm{QC} = \mathrm{VMO} \cap L^\infty$, the space of quasicontinuous functions on ∂D. It can be defined, alternatively, as the largest C^*-subalgebra of the algebra $H^\infty + C$. If C_b^* maps BMO into VMO then, because it also maps L^∞ into itself, it maps L^∞ into QC. Conversely, if C_b^* maps L^∞ into QC then, because it commutes with the conjugation operator, it maps BMO into VMO. Hence, the inclusion $C_b^* L^\infty \subset \mathrm{QC}$ is equivalent to the weak compactness, and thus to the compactness, of C_b on H_1.

J. H. Shapiro and C. Sundberg [6] have recently proved that the compactness of C_b on H_1 implies the compactness of C_b on L^1. In [5] it is proved that the compactness of C_b on L^1 is equivalent to the inclusion $C_b^* L^\infty \subset C$. Combining these results with the theorem proved above, one obtains the following consequence.

COROLLARY. *If* $C_b^* L^\infty \subset \mathrm{QC}$ *then* $C_b^* L^\infty \subset C$.

It would be of interest to have a more direct proof of this corollary, one that better illuminates its function-theoretic meaning.

REFERENCES

1. N. Dunford and J. T. Schwartz, "Linear Operators, Part I," Interscience, New York, 1958.
2. J. García–Cuerva and J. L. Rubio de Francia, "Weighted Norm Inequalities and Related Topics," North–Holland, Amsterdam, 1985.
3. K. Hoffman, "Banach Spaces of Analytic Functions," Prentice–Hall, Englewood Cliffs, NJ, 1962.
4. J. V. Ryff, *Subordinate H^p functions*, Duke Math. J. **33** (1966), 347–354.

5. D. Sarason, *Composition operators as integral operators*, Analysis and Partial Differential Equations, a collection of papers dedicated to M. Cotlar, Marcel Dekker, New York, 1990.
6. J. H. Shapiro and C. Sundberg, *Compact composition operators on L^1*, Proc. Amer. Math. Soc., forthcoming.

Department of Mathematics, University of California, Berkeley, CA 94720

NONEXPANSIVE MAPPINGS AND PROXIMINALITY IN NORMED ALMOST LINEAR SPACES

GEETHA S. RAO

AND

T.L. BHASKARAMURTHI

DEDICATED TO THE MEMORY OF U.N. SINGH

INTRODUCTION:

The concepts of normed almost linear spaces (nals) and Strong normed alomost linear spaces (Snals), which generalize normed linear spaces (nls), were first introduced by G.Godini [2]. In [2] it was proved that nals [Snals] constitute the natural framework for the theory of best simultaneous approximation, by showing that this theory is a particular case of the theory of best approximation in a nals [Snals]. The dual space of a nals X, where the functionals are not linear but "almost linear", was also introduced in [2] to support the idea that the nals is a good concept. A similar theorem of the type of Hahn-Banach theorem, which is an important tool in the theory of normed linear space, is no longer possible in nals. But one of the important corollaries to Hahn-Banach theorem viz. for each $x \in X$, a normed almost linear space, there exists an $f \in X^*$ such that $|||f||| = 1$ and $f(x) = |||x|||$ was given by Godini[3].

Key words: Nonexpansive mappings, Fixed points, Proximinality, Normed alomost linear space.

AMS Classification: 41 A65.

In this paper, the results contained in [6] have been extended in the framework of nals. It was proved in [6] that the set of fixed points of a nonexpansive mapping P:E -> E admits Chebyshev centers, when E is an (AL)-space and the mapping P is a positive one. Theorem 2.11 is a generalization of this result in a certain sense.

1. DEFINITIONS AND NOTATION

1.1. An almost linear space (als) is a set X together with two mappings $s : X \times X \to X$ and $m : R \times X \to X$ satisfying (L_1)-(L_8) below. For $x,y \in X$ and $\lambda \in R$ denote $s(x,y)$ by $x+y$, $m(\lambda,x)$ by λx and $-1x$ by $-x$ and in the sequel $x-y$ stands for $x+(-y)$. Let $x,y,z \in X$ and $\lambda, \mu \in R$. (L_1) $(x+y)+z = x+(y+z)$; (L_2) $x+y = y+x$; (L_3) There exists an element $0 \in X$ such that $x+0 = x$ for each $x \in X$; (L_4) $1x = x$; (L_5) $0x = 0$; (L_6) $\lambda(x+y) = \lambda x+\lambda y$; (L_7) $\lambda(\mu x) = (\lambda\mu)x$; (L_8) $(\lambda+\mu)x = \lambda x+\mu x$ for $\lambda \geq 0$, $\mu \geq 0$.

1.2. (a)A nonempty set Y of an als X is called an almost linear subspace of X, if for each $y_1,y_2 \in Y$ and $\lambda \in R$, $s(y_1,y_2) \in Y$ and $m(y_1,y_2) \in Y$.

(b) An almost linear subspace Y of X is called a linear subspace of X if $s : X \times Y \to Y$ and $m : R \times Y \to Y$ satisfy all the axioms of als.

1.3. For an als X, consider the following sets :

$$V_X = \{ x \in X : x-x = 0 \}$$

$$W_X = \{ x \in X : x = -x \}.$$

Note that the set V_X is a linear subspace of X, and it is the largest one. The als X is ls, if and only if $V_X = X$. On the other hand, the set W_X is an almost linear subspace of X, $W_X = \{ x-x : x \in X \}$ and $V_X \cap W_X = \{0\}$. The als X is a ls, if and only if $W_X = 0$.

1.4. A norm on an als X is a functional

$|||.|||: X \to R$ satisfying (N_1-N_4) below. Let $x,y,z \in X$ and $\lambda \in R$. (N_1) $|||x-z||| \le |||x-y||| + |||y-z|||$; ($N_2$) $|||\lambda x||| = |\lambda|\ |||x|||$, ($N_3$) $|||x||| = 0$ if and only if $x = 0$. By (N_1), it follows that

$$|||x+y||| \le |||x||| + |||y|||.$$

By 1.3, V_X is a ls and so (V_X, $|||.|||$) is a nls. Therefore the weak convergence (denoted by $\rightharpoonup$) can be defined in V_X.

(N4) If $\{v_n\}$ is a net in V_X, $v \in V_X$,

$v_n \rightharpoonup v$, then for each $x \in X$, $|||x-v||| \le \liminf |||x-v_n|||$.
An als X together with $|||.|||: X \to R$ satisfying (N_1)-(N_4) is called a normed almost linear space (nals).

1.5. Let X be an als. a) A functional $f: X \to R$ is called an almost linear functional if (F1)-(F3) hold :

(F_1)	$f(x+y) = f(x)+f(y)$	$(x,y \in X)$
(F_2)	$f(\lambda x) = \lambda f(x)$	$(\lambda \in R,\ \lambda \ge 0,\ x \in X)$
(F_3)	$-f(-x) \le f(x)$	$(x \in X)$.

b) A functional $f: X \to R$ is called a linear functional if (F_1) and (F_2) hold for all $x,y \in X$ and $\lambda \in R$.

1.6. Let $X^\#$ be the set of all almost linear functionals defined on an als X. Then $X^\#$ is an als, for example, see [2]. When X is a nals, for $f \in X^\#$ define

$$|||f||| = \text{Sup}\ \{\ |f(x)| : x \in B_X\ \}, \tag{1}$$

where $B_X = B_X(0,1) = \{y \in X : |||y||| \le 1\}$.

Let $X^* = \{f \in X^\# : |||f||| < \infty\}$.

X^* together with $|||.|||$ defined in (1) is nals. The space together with $|||.|||$ defined in (1) is called the dual space of the nals X. Note that if X is a nls, then the dual space in the preceding definition is the usual dual space of X.

REMARK 1.7. A natural example for nals is furnished by the convex bounded, nonempty subsets of a nls E [2]. Several examples for nals are given by Godini [2].

1.8. Let E be a normed linear space over the real field R and G a nonempty subset of E. For a bounded set $A \subset E$ let

$$\mathrm{rad}_G(A) = \inf_{g\in G} \sup_{a\in A} ||a-g||$$

$$\mathrm{cent}_G(A) = \{g_o \in G : \mathrm{Sup}_{a\in A} ||a-g_o|| = \mathrm{rad}_G(A)\}.$$

The number $\mathrm{rad}_G(A)$ is called the Chebyshev radius of A with respect to G, and an element $g_o \in \mathrm{cent}_G(A)$ is called a Chebyshev centre of A with respect to G. When A is a singleton, say $A = \{x\}$, $x \in E$, then $\mathrm{cent}_G(A)$ is the set of all best approximations of x out of G, denoted by $P_G(x)$ and defined by

$$P_G(x) = \{ g_o \in G : ||x-g_o|| = \inf_{g\in G} ||x-g|| \}. \tag{2}$$

$\mathrm{rad}_G(A)$ is the distance of x to G denoted by dist (x,G) and denoted by

$$\mathrm{dist}\ (x,G) = \inf_{g\in G} ||x-g||$$

It is well known that for any bounded set $A \subset E$,

$$\mathrm{rad}_G(A) = \mathrm{rad}_G(\mathrm{co}\ A) = \mathrm{rad}_G(\bar{A}),$$

$$\mathrm{cent}_G(A) = \mathrm{cent}_G(\mathrm{co}\ A) = \mathrm{cent}_G(\bar{A}),$$

where co stands for the convex hull.

1.9. Let X be a nals, G a subset of X, and $x \in X$. Define dist (x,G) and $P_G(x)$ by (2) and (3), replacing $||\ .\ ||$ by $|||.|||$ and retaining the same definitions for proximinal and Chebyshev sets in a nls. It was observed in [2] that in view of the example 3.2 of [2], if $x \in X$ stands for the bounded, convex, nonempty set $A \subset E$, then for any $G \subset V_X$ (=E), it follows that

$$\text{dist}\ (x,G) = \text{rad}_G(A)$$

$$P_G(x) = \text{cent}_G(A)$$

Consequently, any information obtained concerning the function dist (x,G) and the set-valued mapping $x \to P_G(x)$, when $G \subset V_X$ and X is a nals, is also valid for the function $\text{rad}_G(.)$ and for the set-valued mapping $A \to \text{cent}_G(A)$, where A is a bounded, nonempty subset of E.

1.10. Let X be a nals. A mapping $T : A \to X$ where $A \subset X$ is some nonempty subset, is nonexpansive if, for each pair x and y in A

$$|||Tx-Ty||| \le |||x-y|||.$$

If $K \subset A$ is such that $T(K) \subset K$, then K is T-invariant or invariant under T.

2. PROXIMINALITY IN NALS

The following proposition was given by Godini in [2].

PROPOSITION 2.1. Let X be a nals and G a boundedly weakly compact subset of V_X. Then G is proximinal in X.

This proposition is employed in the proofs of some of the results provided in this paper.

PROPOSITION 2.2. Let $T : X \to X$ be a nonexpansive mapping and $M = \{ x \in X : Tx = x \}$. Let $x \in M$. If $G \subset V_X$ is a T-invariant nonempty subset, then $P_G(x)$ is T-invariant.

PROOF. Let $g_o \in P_G(x)$. Since G is T-invariant, $Tg_o \in G$. Now

$$|||x-Tg_o||| = |||Tx-Tg_o||| \le |||x-g_o||| \le |||x-g||| \quad (g \in G)$$

and this implies that

$$Tg_o \in P_G(x).$$

REMARK 2.3. Under the hypothesis of Proposition 2.1, it follows that

$$T(P_G(x)) \subset P_{T(G)}(x).$$

PROPOSITION 2.4. Let $T : X \to X$ be an idempotent nonexpansive mapping. If $V_X \subset X$ is proximinal and $G \subset V_X$ is range of the idempotent nonexpansive mapping $T|V_X$, then G is proximinal in the range of T.

PROOF. Since $T^2 = T$, $G = T(V_X) = \{ v \in V_X : T(v) = v \}$.

By remark 2.3, it follows that

$$T(P_{V_X}(x)) \subset P_{T(V_X)}(x) = P_G(x).$$

Since $P_{V_X}(x) \neq \phi$, $P_G(x) \neq \phi$. hence G is proximinal in the range of T.

COROLLARY 2.5. Let $T : X \to X$ be an idempotent nonexpansive mapping. If V_X is proximinal in X and $G \subset V_X$ is a norm-one complemented subspace, then G is proximinal in the range of T.

PROOF. If G is norm-one complemented in V_X, then G is the range of the nonexpansive linear projection $T_1 : V_X \to V_X$.
But $T|V_x = T_1$. Hence $T(P_{V_X}(x)) \subset P_{T(V_X)}(x) = P_{T_1(V_X)}(x) = P_G(X)$ and therefore G is proximinal in the range of T.

COROLLARY 2.6. .Let X and X^* be normed almost linear spaces. Let $T : X^{**} \to X^{**}$ be an idempotent nonexpansive mapping. If V_X is norm-one complemented in $(V_X)^{**}$ and $(V_X)^{**}$ is boundedly weakly compact in $V_X{**}$, then V_X is proximinal in the range of T.

PROOF. Since $(V_X)^{**} \subset V_X{**} \subset X^{**}$, and $(V_X)^{**}$ is boundedly weakly compact in $V_X{**}$, by Proposition 2.1, it follows that $(V_x{**})$ is proximinal in X^{**}. Therefore, by Corollary 2.5, V_X is proximinal in the range of T.

REMARK 2.7. Corollary 2.5 is an extension of the well-known result given in Holmes [4], p.184. Corollary 2.6 is an extension of Garkavi's result, Theorem III of [1].

REMARK 2.8. In the proof of Proposition 2.4, the fact that the range of an idempotent mapping is the set of its fixed points was used. The question arises as to whether the set of fixed points of a nonexpansive mapping $T|V_X$ is proximinal, if V_X is proximinal in the range of T.

The solution to this problem in two different cases is provided next. The cases are : a) V_X is reflexive and $T : X \to X$ is a linear idempotent nonexpansive mapping b) V_X is an (AL)-space and $T : X \to X$ is a positive linear idempotent nonexpansive mapping.

THEOREM 2.9. Let $T : X \to X$ be a linear idempotent nonexpansive mapping. Let G be the set of fixed points of $T|V_X$. If V_X is reflexive, then G is proximinal in the range of T.

PROOF. Since $G = \{ v \in V_X : Tv = v \}$, it follows that G is a closed subspace of V_X. Since V_X is reflexive, G is also a reflexive subspace of V_X. By proposition 2.1, G is proximinal in the range of T.

LEMMA 2.10. Let X be a normed almost linear space.
Let $T : X^{**} \to X^{**}$ be a positive idempotent nonexpansive mapping. If $(V_X)^{**}$ is boundedly weakly compact in V_X^{**} and V_X is an (AL)-space, then V_X is proximinal in the range of T.

PROOF. Every (AL)-space, is an injective Banach lattice [5]. From this it follows that V_X is norm-one complemented in $(V_X)^{**}$ [6]. Then by Corollary 2.6, it follows that V_X is proximinal in the range of T.

THEOREM 2.11. Let X be a normed almost linear space.

Let $T_1 : X^{**} \to X^{**}$ be a positive idempotent nonexpansive mapping. If $(V_X)^{**}$ is boundedly weakly compact in V_X^{**} and V_X is an (AL)-space and $T : X \to X$ be a positive linear idempotent nonexpansive mapping, then the set of fixed points of $T|V_X$ is proximinal in the range of T.

PROOF. Let $M = \{ v \in V_X : T_v = v \}$. Clearly, M is a closed subspace of V_X. By using the techniques of [6], it can be shown that M is a sublattice of V_X. Since every closed vector sublattice of an (AL)-space is norm-one complemented in it [6], it follows that M is norm-one complemented in V_X. By lemma 2.10 and Corollary 2.5, M is proximinal in the range of T.

REFERENCES

1. Garkavi, A.L., The best possible net and the best possible cross-section of a set in a normed space, Amer.Math.Soc.Translations (series 2) 39 (1964), 111-132.

2. Godini, G., A framework for Best Simultaneous Approximation: Normed Almost Linear Spaces, J. Approx. Theory 43, 338-358 (1985).

3. Godini, G., Operators in normed almost linear spaces, Rend.Cire.Mat.Palermo, 14, 309-328, 1987.

4. Holmes, R.B. A course on optimization and best approximation, lecture Notes in Mathematics No. 257, Springer-Verlag, Berlin, Heidelberg and New York, 1972.

5. Lotz, H.P., Extentions and liftings of positive linear mappings of Banach lattices, Trans.Amer.Math.Soc.211 (1975), 85-100.

6. Prolla, J.B., Nonexpansive mappings and Chebyshev centres in (AL)-spaces. Approximation Theory IV, Academic Press, New York/London, 1983.

RAMANUJAN INSTITUTE
UNIVERSITY OF MADRAS
MADRAS - 600 005
INDIA

Continuity of Seminorms

Henry Helson and John E. McCarthy

in memory of U.N. Singh

A lower-semicontinuous seminorm on a Banach space is continuous. This result is well known [2] and is proved by a category argument. Banach knew that a Borel seminorm on a Banach space is continuous. A simple proof of this result is given in [3], using measure theory instead of category. The purpose of this note is to extend the theorem and proof to topological vector spaces that are F-spaces. Then is a corollary the Banach-Steinhaus theorem follows for such spaces. We do not claim that these results are new, but we think the proof is simpler than other proofs.

An F-space is a topological vector space whose topology is given by a complete translation-invariant metric.

Theorem. *A Borel seminorm on an F-space is continuous.*

Let q be a Borel seminorm on X, a vector space with complete invariant metric ρ. If q is not continuous, find a sequence (x_n) in X such that x_n tends to 0, but $q(x_n) > \delta > 0$. By passing to a subsequence we can have $\rho(x_n, 0) < 4^{-n}$ for each n. Let $y_n = 2^n x_n$. Then

$$(1) \qquad \rho(y_n, 0) \leq \rho(2^n x_n, (2^n-1)x_n) + \rho((2^n-1)x_n-(2^n-2)x_n) + \ldots + \rho(x_n, 0) \leq 2^n \rho(x_n, 0) < 2^{-n}.$$

At the same time we have

$$(2) \qquad q(y_n) = q(2^n x_n) = 2^n q(x_n) > 2^n \delta.$$

Now the proof from [3] can be replayed. For each sequence $r = (r_1, r_2, \ldots)$ of zeros and ones, define

$$(3) \qquad h(r) = \sum_1^\infty r_n y_n,$$

a series whose partial sums form a Cauchy sequence in X by (1), and which therefore converges. Form the compact abelian group K whose elements are all sequences r with addition in each component modulo 2. The partial sums in (3) are continuous functions from K into X, and convergence is uniform; therefore h is continuous. Hence $k(r) = q(h(r))$ is a Borel function from K to the real line R.

For a given positive number A, let E be the set of r where $k(r) \leq A$. E is a

Borel set, and if A is large enough it has positive measure, because h is finite everywhere. We fix such an A.

The theorem of Steinhaus asserts that E-E (the set of all r-s where r, s belong to E) contains a neighborhood of 0 in K. The definition of the topology in K shows that every neighborhood of 0 contains some set V_k, consisting of all r such that $r_1 = r_2 = \ldots = r_k = 0$. In particular, for every $n > k$, E-E contains the point r^n whose components are 0 except in the nth place, where there is a 1. Choose such an n, and let r and s be points of E such that r-$s = r^n$.

r and s have the same components except in the nth place, where they are different. We reverse r and s if necessary (noting that r-$s = s$-r) so tha' $r_n = 1$, $s_n = 0$. Then

$$y_n = h(r-s) = h(r) - h(s). \tag{4}$$

It follows that

$$q(y_n) = q(h(r)-h(s)) \leq q(h(r)) + q(h(s)) \leq 2A. \tag{5}$$

This has been proved for any $n > k$, so (5) is in contradiction with (2), and the proof is finished.

For completeness, we recall the proof of Steinhaus' theorem. Let g be the characteristic function of the set E of positive measure in K. The function

$$\int_K g(r)g(r+s)\, dr \tag{6}$$

(where dr is Haar measure on K) is a continuous function of s on K, positive for $s = 0$, and therefore positive for all s in a neighborhood V of 0. For each such s, the integrand is positive for some r. That is, if s is in V, then r and $r+s$ are both in E for some r. Hence $s = (r+s)$-r is in E-E as was to be proved.

Here is a version of the theorem of Banach and Steinhaus: *If $\mathfrak{T} = (T)$ is a family of continuous linear operators from the F-space X to a normed vector space Y such that $\|Tx\|$ is bounded for each fixed x in X as T varies over $\mathfrak{T}$, then $\|Tx\|$ is bounded as T varies over $\mathfrak{T}$ and x over some neighborhood of* 0.

It is easy to see that it suffices to prove the result when $\mathfrak{T}$ is a countable family, using the fact that X has a countable neighborhood base at 0. Define $q(x) = \sup_T \|Tx\|$, a seminorm on X. For fixed T the seminorm is continuous, and the supremum is a Borel function because we have assumed $\mathfrak{T}$ to be countable. The theorem just proved shows that q is continuous, and it follows that q is bounded on a neighborhood of 0. This completes the proof.

If the operators T are assumed merely to be Borel, the proof applies and they are in fact continuous.

It would be interesting to find a similar proof for the Open Mapping Theorem (which is easily equivalent to the Closed Graph Theorem). Let T be a continuous one-one mapping of a Banach space X onto a Banach space Y. We would like to show that T^{-1} is continuous. A norm is defined in Y by setting $\|y\|' = \|T^{-1}y\|$, where the second norm is the given norm in X. We have to prove that the primed norm is continuous on Y, and we have shown that this is the case if it is a Borel norm. If X is separable, then by Souslin's theorem [1, p. 67] T is a Borel isomorphism and the result follows. It does not seem easy to remove the hypothesis of sepa-ability, nor is Souslin's theorem an elementary result. `hus our method does not seem advantageous for proving the Open Mapping Theorem itself, but it might be useful in some related subject matter.

References

1. W. A. Arveson, *An Invitation to C^*-Algebras*, Springer-Verlag, 1976.

2. R. E. Edwards, *Functional Analysis*, Holt, Rinehart and Winston, 1965.

3. H. Helson, Boundedness from measure theory. II, in *Operator Theory: Advances and Applications*, Vol. 11, Birkhäuser Verlag, 1983, 179-190.

University of California, Berkeley, CA 94720
Indiana University, Bloomington, IN 47405

Maximal ideals in local Carleman algebras

JAMIL A. SIDDIQI

Abstract. In this paper,we construct characteristic functions for local Carleman classes and use them to determine maximal ideals in local Carleman algebras.

1.Introduction.

Let X be a locally compact Hausdorff space. An algebra $\mathcal{A}(X)$ of complex-valued continuous functions defined on X is called a *standard function algebra*(see[8]) if the following conditions are satisfied:

(S1) $f \in \mathcal{A}(X), f(a) \neq 0 (a \in X)$ implies that there exists a $g \in \mathcal{A}(X)$ such that $g(x) = (1/f(x))$ for all x in some neighbourhood of a.

(S2) For any closed subset E of X and for any a in $X - E$, there is an $f \in \mathcal{A}(X)$ such that $f(E) = 0$ and $f(a) \neq 0$.

Since the descriptions of the ideals in these algebras is well-known, it seems natural to search for necessary and sufficient condi :ons that the Carleman algebras

$$C_M(I) = \left\{ f \in C^\infty(I) : \|f^{(n)}\|_\infty \leq A_f B_f^n M_n \quad (n \geq 0) \right\},$$

where $M = \{M_n > 0\}$ and $I \subseteq \mathbf{R}$ is an interval and the local Carleman algebras

$$C_M^*(I) = \{ f \in C^\infty(I) : f \in C_M \quad \text{locally} \},$$

be standard.

The solution of this problem depends largely on the explicit availability of characteristic functions of these classes for each type of linear interval I. By definition, these are the functions which belong to a class $C_M(I)$ or $C_M^*(I)$ but to no class $C_N(I)$ or $C_N^*(I)$ with $N_n^{\frac{1}{n}} = o(M_n^{\frac{1}{n}}) \quad (n \to \infty)$. Such functions were effectively constructed by T.Bang [1] and others for $I = \mathbf{R}$ and $I = \mathbf{R}_+$ respectively. For local Carleman classes defined on finite intervals, the characteristic functions were given by H.Cartan [2]. These functions are also useful in solving the problems of non-triviality, equivalence, differentiability, analyitcity and superposition for these classes (see [6],[10],[11]).

For local Carleman classes defined on infinite open or semi-closed intervals, the characteristic functions have not been effectively constructed. In this paper, we construct such functions and use them to solve the problem in hand briefly indicating the extensions of classical results mentioned above.

Given any sequence $M = \{M_n\}$ of positive numbers, we set

$$S(r) = \max_{n \leq r} \frac{r^n}{M_n} \quad (r \geq 1)$$

and

$$M_n^o = \max_{r \geq n} \frac{r^n}{S(r)} \quad (n \geq 1). \tag{1}$$

Similarly we set

$$U(r) = \max_{n \leq r} \frac{r^{2n}}{n^n M_n} \quad (r \geq 1)$$

and

$$n^n M_n^f = \max_{r \geq n} \frac{r^{2n}}{U(r)} \quad (n \geq 1). \tag{2}$$

The following regularization theorem due to H.Cartan and S.Mandelbrojt [5] shows that the local classes $C_M^*(I)$ can be alternately described in terms of the sequences M^o and M^f.

Research supported by a NSERC of Canada. Dedicated to the memory of U.N.Singh.

THEOREM A. *Let $M = \{M_n\}$ be a sequence of positive numbers.*

a) For any open interval I, $C^*_M(I) \equiv C^*_{M^o}(I)$.

b) For any interval I, $C^*_M(I) \equiv C^*_{M^f}(I)$.

2.Characteristic functions on arbitrary open intervals.

We first consider the case when I is an infinite open interval. Since a linear transformation of the interval does not change the class, we can ,without loss of generality, take I to be $\mathbf{R}$ or $\mathbf{R}^o_+$.

It is easily seen from (1) that there exist integers $h_n \geq n$ such that

$$\frac{h_n^n}{S(h_n)} \geq \frac{M_n^o}{e}. \tag{3}$$

Let

$$f(x) = \sum_{p=1}^{\infty} \frac{Z_{h_p^2}(\alpha x/h_p)}{2^p S(h_p)},$$

where $0 < \alpha < 1$ and

$$Z_n(x) = \frac{1}{2}[(-1)^{[n/2]}T_n(x) + (-1)^{[(n-1)/2]}T_{n-1}(x)],$$

$T_n(x)$ denoting, as usual, the Chebyshev polynomial of degree n and $[t]$,the integral part of t.

Let $x \in [-\lambda, \lambda]$. Since

$$f^{(n)}(x) = \sum_{p=1}^{\infty} \frac{\alpha^n Z^{(n)}_{h_p^2}(\alpha x/h_p)}{2^p h_p^n S(h_p)},$$

the last sum can be split into two sums namely, Σ_1 and Σ_2, where the first sum runs over all p's with $\lambda^2 \geq h_p^2 \geq n$ and the second over all the p's with $h_p^2 \geq n$ and $h_p > \lambda$.

Since for every $x \in [-1, 1]$ (see [6])

$$|Z_n^{(k)}(\alpha x)| \leq K_\alpha^k n^k, \tag{4}$$

where K_α depends on α, it follows that outside the interval [-1,1]

$$|Z_n^{(k)}(\alpha x)| \leq K_\alpha^k n^k T_{n-k}(x). \tag{5}$$

If we choose $L > 1$, then for each $x \in [-L, L]$

$$|Z_n^{(k)}(\alpha x)| \leq K_\alpha^k n^k (2L)^{n-k}. \tag{6}$$

For $x \in [-\lambda, \lambda]$, using (6),we get

$$|\Sigma_1| \leq \sum_{n \leq h_p^2 \leq \lambda^2} \frac{\alpha^n K_\alpha^n h_p^{2n}}{2^p h_p^n S(h_p)} (2\lambda)^{h_p^2 - n} \leq A_\lambda K_\alpha^n M_n^o$$

while, using (4), we get

$$|\Sigma_2| \leq \sum_{n \leq h_p^2, \lambda < h_p} \frac{\alpha^n K_\alpha^n h_p^{2n}}{2^p h_p^n S(h_p)} \leq K_\alpha^n M_n^o.$$

Thus for each $x \in [-\lambda, \lambda]$,

$$|f^{(n)}(x)| \leq A_\lambda K_\alpha^n M_n^o \qquad (n \geq 0)$$

and $f \in C_M^*(\mathbf{R})$.

Since (see [6])

$$(-1)^{[(k/2)]} Z_n^{(k)}(0) > (\frac{n}{e})^k,$$

it follows from (3) and (7) that

$$\begin{aligned}(-1)^{[\frac{n}{2}]} f^{(n)}(0) &= \sum_{h_p^2 \geq n} \frac{\alpha^n (-1)^{[(n/2)]} Z_{h_p^2}^{(n)}(0)}{2^p h_p^n S(h_p)} \\ &> \frac{\alpha^n h_n^n}{(2e)^n S(h_n)} \\ &\geq \beta^{n+1} M_n^o.\end{aligned}$$

for some $0 < \beta < 1$. Thus

$$|f^{(n)}(0)| \geq \beta^{n+1} M_n^o. \tag{8}$$

We have thus shown that f is a characteristic function of the class $C_M^*(\mathbf{R}) \equiv C_{M^o}^*(\mathbf{R})$. It is also the characteristic function of the class $C_M^*(\mathbf{R}_+^o) \equiv C_{M^o}^*(\mathbf{R}_+^o)$ and indeed of any class $C_M^*(I)$,where $I =]-\lambda, \lambda[$. Thus every local Carleman class defined on an open interval possesses a characteristic function.

3.Characteristic functions on closed or semi-closed intervals.

The construction of the characteristic function for classes $C_M^*(I)$, when I is a closed or a finite or infinite semi-closed interval is similar although the details are different.

It follows from (2) that for each integer n, there exists an integer $k_n \geq n$ such that

$$\frac{k_n^{2n}}{U(k_n)} \geq \frac{1}{e} n^n M_n^f. \tag{9}$$

Set

$$f(x) = \sum_{p=1}^{\infty} \frac{T_{k_p^2}(-(x/k_p^2) + 1)}{2^p U(k_p)}.$$

Let $x \in [0, \lambda]$. Since

$$f^{(n)}(x) = (-1)^n \sum_{k_p \geq \sqrt{n}} \frac{T_{k_p^2}^{(n)}(-(x/k_p^2)) + 1)}{2^p k_p^{2n} U(k_p)}.$$

as above,this sum can be split into two sums,namely, Σ_1 and Σ_2, where the first sum runs over all p's with $\sqrt{n} \leq k_p \leq \sqrt{(\lambda/2)}$ and the second over all p's with $k_p \geq \sqrt{n}, p \geq \sqrt{(\lambda/2)}$.

It is well-known (see [6]) that for each $x \in [-1, 1]$ and for $0 \leq k \leq n$

$$|T_n^{(k)}(x)| \leq \left(\frac{en^2}{2k}\right)^k \tag{10}$$

so that for if $L > 1$, then for $x \in [-L.L]$

$$|T_n^{(k)}(x)| \leq \left(\frac{en^2}{2k}\right)^k (2L)^{n-k}. \tag{11}$$

For $x \in [0, \lambda]$, using (11), we get

$$|\Sigma_1| \leq \sum_{\sqrt{n} \leq k_p \leq \sqrt{(\lambda/2)}} \frac{k_p^{-2n} e^n k_p^{2n} (2\lambda)^{k_p^2 - n}}{2^p U(k_p)(2n)^n} \leq A_\lambda (\frac{e}{2})^n M_n^f$$

while, using (10), we get

$$|\Sigma_2| \leq \sum_{\sqrt{n} \leq k_p, \sqrt{(\frac{\lambda}{2})} \leq k_p} \frac{k_p^{(2n)} e^n}{2^{n+p} n^n U(k_p)} \leq (\frac{e}{2})^n M_n^f$$

and $f \in C_{M^f}(\mathbf{R}_+)$.

Since (see [6])

$$T_n^{(k)}(1) \geq \left(\frac{n^2}{2ek}\right)^k,$$

it follows from (9) that

$$|f^{(n)}(0)| \geq \frac{k_n^{2n}}{4en^n U(k_n)} \geq \left(\frac{1}{4e}\right)^{n+1} M_n^f.$$

Thus f is a characteristic function of the class $C_M^*(\mathbf{R}_+) \equiv C_{M^f}^*(\mathbf{R}_+)$ and indeed of any class $C_M^*(\mathbf{I}) \equiv C_{M^f}^*(\mathbf{I})$ with $I = [0, \lambda[$ or $[0, \lambda]$. We,therefore, conclude that every local Carleman class possesses a characteristic function.

4.Equivalence, differentiability and analyticity of classes.

The existence of characteristic functions for classes $C_M^*(I)$,where I is an arbitrary finite or infinite interval,gives us a simple method to prove the following strengthened version of a classical theorem of H.Cartan and S. Mandelbrojt [4] obtained by functional analytical techniques in [10].

THEOREM 1.

a) If I is an open interval (finite or infinite), then $C_M^*(I) \subseteq C_N^*(I)$ if and only if $(M_n^o)^{\frac{1}{n}} = O\left[(N_n^o)^{\frac{1}{n}}\right] \quad (n \to \infty)$.

b) If I is not an open interval, then $C_M^*(I) \subseteq C_N^*(I)$ if and only if $(M_n^f)^{\frac{1}{n}} = O\left[(N_n^f)^{\frac{1}{n}}\right]$ $(n \to \infty)$.

PROOF: Since $C_M^*(I) \equiv C_{M^o}^*(I)$ when I is an open interval and $C_M^*(I) \equiv C_{M^f}^*(I)$ in every case,the sufficiency of the condition in (1) and (2) is obvious. We prove the necessity of the condition when $I = \mathbf{R}$; other cases can be treated similarly. If $f \in C_M^*(\mathbf{R}) \equiv C_{M^o}^*(\mathbf{R})$ is the characteristic function as constructed above, then

$$\mu^n M_n^o \leq |f^{(n)}(0)|.$$

But, since this function belongs to $C_N^*(\mathbf{R}) \equiv C_{N^o}^*(\mathbf{R})$,

$$|f^{(n)}(x)| \leq A_\lambda B_\lambda^n N_n^o,$$

for each $x \in [-\lambda, \lambda]$. Thus

$$\mu^n M_n^o \leq A_\lambda B_\lambda^n N_n^o \quad (n \geq 1).$$

In other words,

$$(M_n^o)^{\frac{1}{n}} = O\left[(N_n^o)^{\frac{1}{n}}\right] \quad (n \to \infty).$$

A Carleman class is said to be *differentiable* if with each function, it also contains its derivative. It is said to be *analytic* if it contains only analytic functions. Using the characteristic functions,we can similarly extend the corresponding theorems of H.Cartan and S.Mandelbrojt [4] on the differentiability and analyticity of local classes proved for finite intervals to those defined on arbitrary intervals. It should be noted that when I is infinite the theorem on the differentiability of classes are not a simple corollary of Theorem 1.

THEOREM 2. *Let $M = \{M_n\}$ be a sequence of positive numbers.*

a) *If I is a finite open inte: val,then $C_M^*(I)$ is differentiable if and only if $(M_{n+1}^o)^{\frac{1}{n}} = O\left[(M_n^o)^{\frac{1}{n}}\right] \quad (n \to \infty)$.*

b) *If I is a closed or semi-closed interval, then $C_M^*(I)$ is differentiable if and only if $\left(M_{n+1}^I\right)^{\frac{1}{n}} = O\left[(M_n^I)^{\frac{1}{n}}\right] \quad (n \to \infty)$.*

THEOREM 3. *Let $M = \{M_n\}$ be a sequence of positive numbers.*

a) *If I is an arbitrary open interval,then $C_M^*(I)$ is analytic if and only if $(M_n^o)^{\frac{1}{n}} = O(n) \quad (n \to \infty)$*

b) *If I is a closed or semi-closed interval,then $C_M^*(I)$ is analytic if and only if $(M_n^I)^{\frac{1}{n}} = O(n) \quad (n \to \infty)$*

5.Inverse-closed local algebras. Unlike Carleman classes defined on $\mathbf{R}$ and $\mathbf{R}_+$, a local Carleman class need not be an algebra. The following theorem gives a necessary and sufficient condition for this to be true for such classes.

THEOREM 4. *The class $C_M^*(I)$ is an algebra if and only if*

$$M_n M_m \leq AB^{n+m} M_{n+m}, \tag{12}$$

PROOF: Suppose (12) holds. Since $C_M^*(I)$ is linear, it is sufficient to prove that if f and g are in $C_M^*(I)$, then so is their product fg. But this follows easily from the Leibnitz formula:

$$(fg)^{(n)}(x) = \sum_0^\infty \binom{n}{k} f^{(n)}(x) g^{(n-k)}(x).$$

To prove that (12) is necessary, we proceed as follows. Suppose first that I is an open interval which, without loss of generality, can be taken to be symmetric with respect to the origin. By Theorem A, $C_M^*(I) \equiv C_{M^o}^*(I)$. Now for any x in a compact subinterval J of I also symmetric with respect to the origin, we have

$$|(fg)^{(n)}(x)| = |\sum_0^\infty \binom{n}{k} f^{(k)}(x) g^{(n-k)}(x)| \leq A_J B_J^n M_n^o.$$

If we choose f to be characteristic function of $C_M^*(I)$ and set $g(x) = f(-x)$, then, as shown above,

$$(-1)^{[j/2]} f^{(j)}(0) \geq 0 \qquad \text{for all} \qquad j \geq 0$$

so that

$$f^{(k)}(0)g^{(n-k)}(0) = (-1)^{n-k+[k/2]+[(n-k)/2]}|f^{(k)}(0)||f^{(n-k)}(0)|$$

It is easily seen that for each $0 \leq k \leq n$

$$n-k+[\frac{k}{2}]+[\frac{n-k}{2}] = \frac{n-1}{2} \quad \text{or} \quad \frac{n}{2}$$

according as n is odd or even. It follows,using (8), that

$$\beta^n M_k^o M_{n-k}^o \leq |f^{(k)}(0)f^{(n-k)}(0)| \leq A_J B_J^n M_n^o.$$

which shows that (12) holds. A similar reasoning can be used to show that (12) holds with $M \equiv M^I$ when I is arbitrary.

THEOREM 5. *If $C_{n!}^*(I) \subseteq C_M^*(I)$,then $C_M^*(I)$ is always an algebra.*

PROOF:) that

Since $C_{n!}^*(I) \subseteq C_M^*(I)$, $C_M^*(I) \equiv C_{M^d}^*(I)$ (see Lemma 1 of [10]) where $M^d \equiv \{M_n^d\}$ with $n^n M_n^d \equiv (n^n M_n)^c$. Since $\{(n^n M_n)^c\}$ is log-convex,

$$\begin{aligned} M_n^d M_m^d &= \frac{(n^n M_n)^c}{n^n}\frac{(m^m M_m)^c}{m^m} \\ &\leq \frac{\left((m+n)^{(m+n)} M_{m+n}\right)^c}{n^n m^m} \\ &= \frac{(m+n)^{m+n}}{n^n m^m} M_{m+n}^d. \end{aligned}$$

But $n! \leq n^n \leq e^n n!$. Hence

$$\frac{(m+n)^{m+n}}{n^n m^m} \leq \frac{(m+n)! e^{m+n}}{n! m!} \leq (2e)^{m+n},$$

and it follows that $C_M^*(I)$ is an algebra.

Following W.Rudin [9], we say that an algebra $C_M^*(I)$ is *inverse-closed* if $f \in C_M^*(I)$ and $\inf_x |f(x)| > 0$ implies that $f^{-1} \in C_M^*(I)$.More generally,analytic functions are said to *operate* on $C_M^*(I)$ if for any $f \in C_M^*(I)$ and for any g analytic in an open set containing the closure of the range of f, $gof \in C_M^*(I)$. A sequence $A \equiv \{A_n\}$ is said to *almost- increasing* if there exists a constant $K > 0$ such that $A_m \leq K A_n$ whenever $m \leq n$. In [11] ,we proved the following theorem that characterizes the inverse-closed Carleman algebras.

THEOREM B. *Let X be a nontrivial local Carleman class. The following propositions are equivalent:*

a) The sequence $A \equiv \{A_n\}$ is almost increasing.

b) Analytic functions operate on X.

c) X is an inverse-closed algebra.

Here $A_n = (M_n^o/n!)^{\frac{1}{n}}$ or $(M_n^I/n!)^{\frac{1}{n}}$ according as I is open or not.

We now give some elementary results concerning the relationship between local and inverse-closed local Carleman algebras.

THEOREM 6. *Every algebra $C^*_M(I)$ is contained in an inverse-closed algebra $C^*_{M^\bullet}(I)$ which is minimal in the following sense: If an algebra $C^*_{M'}(I)$ contains $C^*_M(I)$ and is an inverse-closed algebra, then $C^*_{M'}(I) \supseteq C^*_{M^\bullet}(I)$.*

PROOF: Let I be an open interval. Put $A^*_n = \max_{s\leq n} A^o_s$, where $A^o_s = (M^o_s/s!)^{\frac{1}{s}}$. Define $M^*_n = n!\,(A^*_n)^n$. Then

$$M^o_n = n!\,(A^o_n)^n \leq n!\,(A^*_n)^n = M^*_n.$$

Since the sequence $\{A^*_n\}$ is increasing, by Theorem B, $C^*_{M^\bullet}(I)$ is inverse-closed. Hence $CalC^*_{M^o}$ is an inverse-closed algebra containing $C^*_{M^o}(I)$. Suppose that $C^*_{M'}(I)$ is an algebra that contains $C^*_{M^\bullet}(I)$. It follows that there exists a constant $\lambda > 0$ such that $M^*_n \leq \lambda^n\,(M'_n)^o \quad (n \geq 0)$. Since $C^*_{M'}(I)$ is inverse-closed,

$$(M'^o_s/s!)^{\frac{1}{s}} \leq K\,(M'^o_n/n!)^{\frac{1}{n}}$$

for all $s \leq n$. It follows that for all $s \leq n$.

$$(M^*_s/s!)^{\frac{1}{s}} \leq \lambda K\,(M'^o_n/n!)^{\frac{1}{n}}$$

or

$$(M^*_n/n!)^{\frac{1}{n}} \leq \lambda K\,(M'^o_n/n!)^{\frac{1}{n}}.$$

and consequently

$$(M^*_n)^{\frac{1}{n}} = O\left[(M'^o_n)^{\frac{1}{n}}\right].$$

Thus $C^*_{M^\bullet}(I) \subseteq C^*_{M'^o}(I) = C^*_{M'}(I)$.

The case when I is not open can be treated analogously.

The following analogue of a theorem of W.Rudin also holds for local Carleman classes.

THEOREM 7. *The intersection of all local Carleman classes that are inverse-closed and non-quasianalytic is precisely the class $C^*_{(n\log n)^n}(I)$.*

PROOF: Let I be open. Since $C^*_M(I)$ is non-quasianalytic, it is not trivial. Since it is also inverse-closed, the sequence $A = \{A_n\}$ with $A_n = (M^o_n/n!)^{\frac{1}{n}}$ is almost increasing. Since this class contains the analytic class, by a theorem of H.Cartan [3] (see also [11]), $C^*_M(I) \equiv C^*_{M^o}(I) \equiv C^*_{M^c}(I)$. Since A is almost increasing it follows that the sequence $\{\alpha_n\}$, where $\alpha_n = (M^c_n)^{\frac{1}{n}}/n$, is also almost increasing. Hence there exists a constant $K > 0$ such that $\alpha_m \leq K\alpha_n$ whenever $m \leq n$. Since, by Denjoy-Carleman theorem (see [6]), the series

$$\sum (M^c_n)^{-\frac{1}{n}}$$

converges, so does the series

$$\sum (n\alpha_n)^{-1}.$$

But

$$\sum_{\sqrt{n}\leq s\leq n} \frac{1}{s\alpha_s} \geq \frac{T}{K\alpha_n}\sum \frac{1}{s} \sim \frac{1}{2K\alpha_n}\log n.$$

The sum on the left tends to zero as $n \to \infty$; hence

$$\lim_{n\to\infty} \frac{\alpha_n}{\log n} = \infty$$

and this means that $C^*_{(n\log n)^n}(I) \subseteq C^*_M(I)$

To prove the other half, we consider a function $f \notin C^*_{(n\log n)^n}(I)$ and construct a non-quasianalytic class $C^*_M(I)$ with M log-convex and $\{(M_n^{\frac{1}{n}}/n)\}\uparrow$ such that $f \notin C^*_M(I)$. This is possible since the intersection of all non-quasianalytic classes is the analytic class. This class is inverse- closed,by Theorem B. The construction of such a class can be done following the lines of Rudin [9] (see pp.805-806). The proof when I is not open is similar. We need only observe that,as proved in [11] ,in this case, we have $C^*_M(I) \equiv C^*_{M^f}(I) \equiv C^*_{M^d}(I)$.

We call a local Carleman algebra $\mathcal{A} = C^*_M(I)$ *weakly inverse-closed* if whenever $f \in$ A $f \neq 0$ then $(1/f) \in \mathcal{A}$. The analysis used to prove Theorem B (see [9]) yields the following theorem:

THEOREM 8. *Let $\mathcal{A} = C^*_M(I)$ be a nontrivial local Carleman class.Then $\mathcal{A}$ is an inverse-closed algebra if and only if $\mathcal{A}$ is weakly inverse-closed algebra.*

PROOF: We need only show that if $\mathcal{A}$ is inverse-closed,it is weakly inverse-closed. In view of Theorem A, our hypothesis implies that the sequence A as defined there is almost increasing. It follows from the formula of Faà di Bruno ,that if $f \in \mathcal{A}$, then locally $f^{-1} \in \mathcal{A}$ (see [11]).

6. Maximal Ideals in local algebras.

The following theorem describes the maximal ideals and complex homomorphisms in the algebra $C^*_M(I)$:

THEOREM 9. *Let $\mathcal{A} = C^*_M(I)$ be inverse-closed. Then every maximal ideal in $\mathcal{A}$ is of the form*

$$\mathcal{I} = \mathcal{I}_x = \{f \in \mathcal{A} : f(x) = 0\}$$

for some $x \in I$ and every complex homomorphism of $\mathcal{A}$ is a point evaluation.

PROOF: Since $\mathcal{A}$ is inverse-closed, $\mathcal{A} \supseteq C^*_{\{n!\}}(I)$. Hence $\mathcal{A}$ is an algebra and $f \in \mathcal{A}$, where $f(x) = e^{-x^2}$. Applying a criterion due to E.Michael (see [7],p.54,Proposition 12.5) with $a_1(x) = f(x)$, we conclude that every complex homomorphism of $\mathcal{A}$ is a point evaluation. If $\mathcal{I}$ is a maximal ideal in $\mathcal{A}$, then $\mathcal{A}/\mathcal{I} \simeq$ C. Let $\psi : \mathcal{A} \to$ C be the canonical mapping defined by the mappings $f \to f + \mathcal{I} \to \lambda$. Then $\psi(f) = f(x_0)$ for every $f \in \mathcal{A}$,where $x_0 \in \mathcal{I}$. But then $\mathcal{I} = \psi^{-1}(0) = \{f \in \mathcal{A} : \psi(f) = f(x_0)\}$. Thus $\mathcal{I} = \mathcal{I}_{x_0}$.

It should be noted that our description of maximal ideals is purely in algebraic sense since we have not endowed $\mathcal{A}$ with a structure of a topological algebra.

7.Standard local algebras.

We recall that a Carleman class or a local Carleman class $\mathcal{A}$ is called *quasianalytic* if $f \in \mathcal{A}$ and $f^{(n)}(a) = 0 \quad (n \geq 0)$ imply that $f \equiv 0$. The following theorem characterizes local classes that are standard:

THEOREM 10. *Let $\mathcal{A} \equiv C^*_M(I)$. Then $\mathcal{A}$ is standard if and only if $\mathcal{A}$ is inverse-closed and non-quasianalytic.*

PROOF: Suppose that $\mathcal{A}$ is standard, y (S2) of the definition,$\mathcal{A}$ is non-quasianalytic. Suppose now that I is an open interval. Since $\mathcal{A}$ contains the analytic class, $C^*_M(I) \equiv C^*_{M^c}(I)$. Let $\{r_n\}$ be a sequence such that

$$\frac{r_n^n}{T(r_n)} = M_n$$

If $a \in I$, the function f defined by setting

$$f(x) = \sum_{j=1}^{\infty} \frac{1}{2^j} \frac{e^{r_j(x-a)}}{T(r_j)}$$

belongs to $\mathcal{A}$ and is such that

$$f^{(n)}(a) = \sum_{j=1}^{\infty} \frac{1}{2^j} \frac{(ir_j)^n}{T(r_j)} = i^n s_n,$$

where

$$s_n \geq \frac{1}{2^n} \frac{r_n^n}{T(r_n)} = \frac{M_n^c}{2^n}.$$

If $|\lambda| > \|f\|_\infty$, $\lambda - f \in \mathcal{A}, \lambda - f(a) \neq 0$ so that by (S1), $(\lambda - f)^{-1}$ belongs to $\mathcal{A}$ locally at a. Applying Faà di Bruno's formula, we conclude that $\left\{(M_n^c/n!)^{\frac{1}{n}}\right\}$ is almost increasing and hence $\mathcal{A}$ is inverse-closed (see [11]).

To prove that the conditions are sufficient,we proceed as follows. Since $\mathcal{A}$ is is non-quasianalytic,the condition (S2) holds. Let $f \in \mathcal{A}$ such that $f(a) \neq 0, (1/f) \in C_{M^c}([a - \eta, a + \eta])$. Such a function exists since $\mathcal{A}$ is non-quasianalytic and that $\left\{(M_n^c/n!)^{\frac{1}{n}}\right\}$ is almost increasing. Since $\mathcal{A}$ is non-quasianalytic,there exists a $g_1 = 1$ in a neighbourhood $U_a \subseteq [a - \eta, a + \eta]$ with support $g_1 \subseteq [a - \eta, a + \eta]$. Then $g = (g_1/f) \in \mathcal{A}, g(x) = (1/f(x))$ for $x \in U_a$. Then (S1) holds. Thus $\mathcal{A}$ is standard.

References

1. T.Bang, "Om quasianalytiske Funktioner," Nyt Nordisk Forlag, Copenhagen, 1949.
2. H.Cartan, *Solution d'une problème de Carleman pour un intervalle fermé fini*, C.R. Acad. Sci. Paris **208** (1939), 716-718.
3. H.Cartan, "Sur les classes de fonctions définies par des inégalités portant sur les dérivées successives," Herman, Paris, 1940.
4. H.Cartan & S. Mandelbrojt, *Solution du problème d'équivalence des classes de fonctions indéfiniment dérivables*, Acta Mathematica **72** (1940), 31-39.
5. A.Mallios, "Topological Algebras," North-Holland Math.Studies vol.124, Amsterdam, 1986.
6. S.Mandelbrojt, "Séies adhérentes,régularisations des suites,applications," Gauthier-Villars, Paris, 1952.
7. E.Michael, "A locally multiplicatively-convex topological algebra," Amer. Math. Soc. Memoir no 11, Providence R.I., 1971.
8. H.Reiter, "Classical Harmonic Analysis and Locally Compact Groups," Oxford Mathematical Monographs, Oxford, 1968.
9. W.Rudin, *Division in algebras of infinitely differentiable functions*, J.Math. Mech. **11** (1962), 797-809.
10. J.A.Siddiqi, *On the equivalence of classes of infinitely differentiable functions* , Izv.Akad.Nauk Armjan. SSR Ser.Mat. **19** (1983), 19-30,333; *ibid*, Soviet J.Contemporary Math. Anal. **19** (1984), 18-29,69.
11. J.A.Siddiqi, *Inverse-closed Carleman algebras*, Proc. Amer. Math. Soc. **109** (1990), 357-367.

Department of Mathematics and Statistics,
Laval University, Québec,Canada,G1K 7P4

Convergence of Polynomials with Restricted Zeros

J.G.ClUNIE

Dedicated to the memory of U.N. Singh

1. INTRODUCTION

Because of the attention it has received in recent years the Laguerre-Polya class of entire functions: *L-P* for short, is quite well known. $f \in L\text{-}P$ if, and only if, it is the local uniform limit in $\mathbb{C}$ of a sequence of polynomials with only real zeros. Such functions are those of the form

$$(1) \qquad f(z) = c.\, z^N \prod_k ((1-(z/z_k)).e^{(z/z_k)}).e^{-az^2+bz} ,$$

where a,b,c, are constants with $a,b \in \mathbb{R}$ and $a \geq o$; N is a non-negative integer; the z_k are non-zero real numbers with $\sum_k z_k^{-2} < \infty$

A partial result is due to Laguerre [3] and the complete result was obtained by Polya [7].

A less well known class of entire functions is the Polya - Obrechkoff class, *P* - O for short. The definition of *P-O* is similar to that for *L-P*, but now the polynomials have their zeros in the closed upper half-plane instead of being restricted to lie on the real axis. Such functions are those of the form(1) with c,a,N as before, but now $z_k \in \mathbb{C}$ with $\text{lim}\, z_k \geq o$

and $\sum_k |z_k|^{-2} < \infty$; $\sum_k \text{lim}\, z_k^{-1}$ converges and $\text{lim}\, b + \sum_k \text{lim}\, z_k^{-1} \geq o$.
The relevant results for this class are due to Polya[8] and Obrechkoff [5], [6].

At a meeting where I gave a talk in which I considered *L-P*, Harold Shapiro asked about uniform convergence in a closed interval of $\mathbb{R}$ of a sequence of polnomials having only real zeros. I could deal with this question, but I felt that the results must be known and I wrote to Jacob Korevaar in Amsterdam asking for information. The results I had obtained were indeed known to a number of people; and Korevaarsent sent me a paper of himself and

Loewner [2] which not only answered the question raised by Shapiro, but coped with the similar question for polynomials having zeros in the closed upper half-plane. From this paper I learned of the class *P-O* that has the same relationship to the latter question as *L-P* does to the former.

The argument of Korevaar and Loewner is based on an extension of a result of Lindwart and Polya [4]. My argument is quite different and is a nice application of normal families [1, §.3] Consequently I hope that the reader will find this paper interesting although it contains nothing new in the way of results.

2. CONVERGENCE IN AN INTERVAL AND *L-P*

THEOREM 2.1 Let (p_n) be a sequence of polynomials with only real zeros and suppose that $p_n(x) \to f(x)$ $(n\to\infty)$ uniformly in $[a,b]$, where $a,b,\in \mathbb{R}$ and $a < b$. Then f is the restriction to $[a,b]$ of a function in *L-P* (again denoted by f).

If $f \neq o$, then (p_n) converges locally uniformly in $\mathbb{C}$ to f.

Before the proof of this theorem I would like to make a couple of remarks. Suppose that $D \subseteq \mathbb{C}$ is a domain and (f_n) is a sequence of functions analytic in D. (f_n) is said to be normal in D if given any compact subset $K \subseteq D$, there is a sequence of (f_n) that converges uniformly on K. However, it is not hard to see that if in addition there is a compact subset $K_o \subseteq D$ and K_o contains infinitely many points and the whole sequence (f_n) converges on K_o, then the whole sequence (f_n) converges uniformly on each compact subset K. In this case one says that (f_n) converges locally uniformly in D.

The proof that follows is presented analytically and it may seem rather complicated in this form. However, if the reader pictures where the various points and neighbourhoods lie in $\mathbb{C}$ he (or she) will realise that the basic ideas are really very simple.

PROOF OF THEOREM 2.1

If $f \equiv o$ the result is trivial. Suppose otherwise and then, since $f(x)$ is continuous on $[a,b]$, there is a closed sub-interval (of positive length) on which $|f(x)| \geq m > 0$ for some constant m. By making a transformation of the kind $x \to \xi x + \eta$ for suitable $\xi, \eta \in \mathbb{R}$ we can, without loss of generality, suppose that $p_n(x) \to f(x)$ $(n \to \infty)$ uniformly on $[-1,1]$ and that $|f(x)| \geq m > 0$ $(-1 \leq x \leq 1)$.

If $\xi \in \mathbb{R}$, $z = x+iy$ (as usual), then

$$|z - \xi| = |(x-\xi) + iy| \geq |x - \xi|$$

and so $|p_n(z)| \geq |p_n(x)|$. Therefore $(1/p_n)_{n \geq n_o}$ for some n_o is bounded in the strip $\{-1 < \text{Re}z < 1\}$ and so is normal in this strip [10,§ 5.22]. Since $1/p_n(x) \to 1/f(x)(n\to\infty)$ in $[-1,1]$, by the first of the above remarks it follows that $(1/p_n)$ converges locally uniformly in the strip to a non-zero analytic function; and hence (p_n) converges locally uniformly in the strip to an analytic function that extends f and which we also denote by f.

At this point we could complete the proof by quoting a result of Polya [7], but instead I shall use an argument that requires once more the idea of a normal family. There are some points of similarity as regards the functions that occur in my argument and that of Polya, but the basic idea is quite different.

We are now concerned with extending the local uniform convergence of (p_n) from the strip $\{-1<\text{Re}z<1\}$ to $\mathbb{C}$. Let $\xi, \xi_o \in \mathbb{R}$ and $Z = X+iY(X,Y \in \mathbb{R})$. Note that

$$|(Z-\xi+\xi_o)(-Z-\xi-\xi_o)| = |Z^2-(\xi_o-\xi)^2|$$
$$\leq |Z^2| + (\xi_o-\xi)^2;$$

and that for $\rho \in \mathbb{R}$,

$$|i\rho-\xi+\xi_o|^2 = \rho^2+(\xi_o-\xi)^2.$$

Hence for $|Z| \le \rho$,

$$|(Z-\xi+\xi_0)(-Z-\xi-\xi_0)| \le |i\rho-\xi+\xi_0|^2.$$

If we regard $Z-\xi$ as a typical factor of $p_n(z)$ it follows that for $|Z| \le \rho$,

$$|p_n(Z+\xi_0).p_n(-Z+\xi_0)| \le |p_n(i\rho+\xi_0)|^2$$

Define

$$Q_n(Z) = p_n(Z+\xi_0).p_n(-Z+\xi_0),$$

where ξ_0 is fixed and satisfies $-1<\xi_0<1$. From the normality of (p_n) in $\{-1<\text{Re}z<1\}$ and the preceding observation it follows that (Q_n) is uniformly bounded on each compact subset of $\mathbb{C}$ and that (Q_n) is normal in $\mathbb{C}$ [10,§5.22]. Therefore (Q_n) converges locally uniformly in $\mathbb{C}$ to a function that coincides with $f(Z+\xi_0).f(-Z+\xi_0)$ for $-1+|\xi_0| < X < 1-|\xi_0|$.

Suppose now that $Z_0 = X_0+iY_0$, where $-1+\xi_0<X_0<1+\xi_0$ and set $z_0 = Z_0+\xi_0$ so that $-1+2\xi_0<x_0<1+2\xi_0$ ($x_0=\text{Re}z_0$). If $\hat{z}_0= -Z_0+\xi_0$, then $-1 < \hat{x}_0 < 1$ ($\hat{x}_0=\text{Re}\hat{z}_0$). In a neighbourhood of $\hat{z}_0$ contained in $\{-1<\text{Re}<1\}$, $(p_n(z))$ will converge uniformly to $f(x)$ and so $(p_n(Z+\xi_0))$ will converge uniformly in the corresponding Z neighbourhood of $-Z_0$ to $f(Z+\xi_0)$. Consequently, in the neighbourhood about Z_0 that is the reflection in the Z-origin of the neighbourhood about $-Z_0$, $(p_n(-Z+\xi_0))$ will converge uniformly to $f(-Z+\xi_0)$ and

$$p_n(Z+\xi_0) = \frac{Q_n(Z)}{p_n(-Z+\xi_0)}$$

will converge uniformly to an analytic function. By varying ξ_0 in $(-1,1)$ we can extend the local uniform convergence of $(p_n(z))$ to the strip $\{-3<\text{Re}Z<3\}$.

By this argument one can step by step extend the local uniform convergence of (p_n) to $\mathbb{C}$ and hence the result of the theorem follows.

3. CONVERGENCE IN AN INTERVAL AND P-O

The following theorem is due to Korevaar and Loewner [2] and the proof I give is essentially the same as the one they give, but the presentation is a little different.

THEOREM 3.1 Let (p_n) be a sequence of polynomials whose zeros lie in the closed upper half-plane. Suppose that $p_n(x) \to f(x) (n \to \infty)$ uniformly on $[a,b]$, where $a,b \in \mathbb{R}$ and $a < b$. Then f is the restriction to $[a,b]$ of a function in P-O (again denoted by f).

If $f \not\equiv o$, then (p_n) converges locally uniformly in $\mathbb{C}$ to f.

PROOF. For $x \in [a,b]$ let

$$p_n(x) = g_n(x) + ih_n(x),$$

$$f(x) = g(x) + ih(x),$$

where g_n, h_n, g, h are real valued on $[a,b]$. It is known [9,p.8] and not difficult to prove, that the polynomials g_n, h_n, have only real zeros. Clearly $g_n(x) \to g(x)$, $h_n(x) \to h(x)$ $(n \to \infty)$ uniformly on $[a,b]$.

Assume that $f \not\equiv o$. If $\alpha, \beta \in \mathbb{R}$, then

$$(\alpha+i\beta)p_n = \alpha g_n - \beta h_n + i(\beta g_n + \alpha h_n),$$

$$(\alpha+i\beta)f = \alpha g - \beta h + i(\beta g + \alpha h),$$

and a suitable choice of α, β will ensure that $\alpha g - \beta h \not\equiv o$, $\beta g + \alpha h \not\equiv o$. From 2.1 it now follows that g, h and hence f extend to entire functions and that as $n \to \infty$,

$$\alpha g_n - \beta h_n \to \alpha g - \beta h,$$

$$\beta g_n + \alpha h_n \to \beta g + \alpha h$$

locally uniformly in $\mathbb{C}$. Therefore (p_n) converges locally uniformly to f in $\mathbb{C}$ and $f \in P\text{-}O$, by definition.

REFERENCES

1. W.K.Hayman, Meromorphic Functions, Oxford (1975).

2. J.Korevaar and C.Loewner, Approximation on an arc by polynomials with restricted zeros, Konink. Neder. Akademie van Wetenschappen - Amsterdam, Proc. Series A, 67(1964), 121-128.

3. E.Laguerre, Sur les fonctions du genre zéro et du genre un, Oeuvres 1 (1898), 174-177.

4. E. Lindwart and G. Polya, Über einen Zusammenhang zwischen der Konvergenz von Polynomfolgen und der Verteilung ihrer Wurzeln, Rend. Circ.Mat.Palermo 37(1914), 297-304.

5. N.Obrechkoff, Sur les racines des équations algébriques, Annuaire de Sofia 23(1927), 177-200.

6. N.Obrechkoff, Quelques classes de fonctions entieres limites de polynomes et de fontions meromorphes limites de fractions rationelles, Act. Sci. et Ind. no. 891(1941), PARIS.

7. G.Polya, Über Annaherung durch Polynome mit lauker reellen Wurzeln, Rend. Circ. Mat. Palermo 36(1913), 279-295.

8. G.Polya, Sur les opérations fonctionnelles linéaires échangeables avec la derivation et sur les zeros des polynomes, C.R.Acad Sci. Paris 183(1926), 413-414.

9. G.Polya and G.Szego, Aufgaben und Lehrsätze aus der Analysis, Berlin (1925).

10. E.C.Titchmarsh, The Theory of Functions, OUP(1964).

Dept. of Maths.,
University of York,
York YO1 5DD, U.K.

ON A THEOREM OF PÓLYA

Jean-Pierre Kahane

dedicated to the memory of U. N. Singh

Pólya developed the theory of entire functions of exponential type in relation with singularities of Taylor and Dirichlet series, a topic which was familiar to the students of S. Mandelbrojt when both U. N. Singh and I worked under his direction in Paris. There is an impressive collection of theorems of Pólya in this area. The theorem which I shall consider (Pólya 1915) has a special flavour through its connection with number theory. Roughly speaking, it expresses that 2^z is the smallest entire function which maps the set of positive integers into itself. According to R. P. Boas, this is "one of the most striking results in analysis".

Several extensions of this theorem were given. We refer to the comments R. P. Boas in the collected works of Pólya [5] and to the article of R. M. Robinson [6] for references and complements.

My purpose is to give a simple approach of Pólya's theorem and of some of its extensions, using an interesting result of Fekete (1926).

Let us begin by a few classical notions on the class of entire functions of exponential type, which we shall denote by E. Each function $f \in E$ can be written (in many ways) in the form

$$f(z) = \int e^{\varsigma z} d\mu(\varsigma) \tag{1}$$

where $d\mu$ is a bounded complex measure with compact support in $\mathbf{C}$, the complex plane. The intersection of the convex hulls of the supports of measures $d\mu$ for which (1) holds is called the conjugate diagram of f and will be denoted by $CD(f)$. Given any open set G which contains $CD(f)$, there exists a measure $d\mu$ such that (1) holds and the support of $d\mu$ lies in G. The exponential type of f, denoted by $ET(f)$, is

$$ET(f) = \sup\{|\varsigma|,\ \varsigma \in CD(f)\}. \tag{2}$$

Given $\theta \in [0, 2\pi]$, the exponential type of f in the direction θ, or indicator function of f at θ, is denoted by $h_f(\theta)$ and defined as

(3) $$h_f(\theta) = \lim_{r\to\infty} \sup r^{-1} \log | f(re^{i\theta}) |$$

so that

(4) $$\begin{cases} h_f(\theta) = \sup\{\mathrm{Re}\, \varsigma e^{i\theta}, \quad \varsigma \in CD(f)\} \\ ET(f) = \sup_\theta h_f(\theta). \end{cases}$$

When $h_f(\theta) = h_f(\pi) = 0$, one says that the real exponential type of f is 0 ; in this case $ET(f)$ equals the imaginary exponential tyne $\sup(h_f(\frac{\pi}{2}), h_f(\frac{3\pi}{2}))$.

For example, if $f(z) = 2^z$, $CD(f) = \{\log 2\}$, $ET(f) = h_f(0) = \log 2$. If $f(z) = 2\cos\frac{\pi}{3}z$, $CD(f) = [-i\frac{\pi}{3}, i\frac{\pi}{3}]$, f has real exponential type 0 and $ET(f) = \frac{\pi}{3}$. If $f(z) = \sin \pi z$, $CD(f) = [-i\pi, i\pi]$, f has real exponential type 0 and $ET(f) = \pi$. All these notions are due to Pólya and can be found in Boas [1].

Let us denote by EI the class of entire functions $f \in E$ which map $\mathbb{N}(= \{0, 1, 2, \cdots\})$ into $\mathbb{Z}(= \{\cdots - 1, 0, 1, \cdots\})$. For example, 2^z, $2\cos\frac{\pi}{3}z$, $\sin \pi z$ belong to EI. What can we say on $f \in EI$ when we know that $CD(f)$ is in a given set, or when we know that $d\mu$ is supported by a given set ?

In order to study this question we can operate a first reduction. Given $d\mu$, let $d\mu^*$ be its image through the map $(x + iy) \to (x + iy^*)$, where $-\pi \le y^* < \pi$ and $y - y^*$ is a multiple of 2π. Then, for each $n \in \mathbb{Z}$,

$$\int e^{n\varsigma} d\mu^*(\varsigma) = \int e^{n\varsigma} d\mu(\varsigma).$$

Writing

$$f^*(z) = \int e^{\varsigma z} d\mu^*(\varsigma)$$

$f^*(z) - f(z)$ is a multiple of $\sin \pi z$, and the conditions $f^* \in EI$ and $f \in EI$ are the same. Therefore, replacing f by f^*, we can restrict ourselves to the case when $CD(f)$ is contained in the strip $| \mathrm{Im}\, z | \le \pi$, that is $h_f(\frac{\pi}{2}) \le \pi$, $h_f(\frac{3\pi}{2}) \le \pi$. Moreover we can suppose that $d\mu$ in (1) is supported by the strip $| \mathrm{Im}\, z | \le \pi$, with no mass on the line $\mathrm{Im}\, z = \pi$. Then the map $\varsigma \to Z = e^\varsigma$ is bijective on the carrier of $d\mu$ and we can write (1) in the form

(5) $$f(z) = \int Z^z d\sigma(Z) \qquad (-\pi \le \arg Z < \pi)$$

$d\sigma$ being the image of $d\pi$ through the map $\varsigma \to Z$.

From now on we suppose (5) and $f \in EI$. Here is the main tool.

LEMMA. *Let $P(Z)$ be a polynomial with integral coefficients such that $|P(Z)| < 1$ on the support of $d\sigma$. Given $\ell \in \mathbb{N}$ we have*

$$\int (P(Z))^k Z^\ell d\sigma(Z) = 0 \tag{6}$$

when k is large enough.

Proof. The first member of (6) is an integer because $f \in EI$ and $P(Z)$ has integral coefficients and it tends to 0 as $k \to \infty$ because $|P(Z)| < 1$ on the support of $d\sigma$.

Whenever we choose $P(Z)$, the lemma provides us with a useful information on EI.

Let us begin with $P(Z) = Z - 1$, and $\ell = 0$. According to (6)

$$\int (Z-1)^k d\sigma(Z) = 0$$

when k is large enough, say $k \geq \kappa$. Therefore

$$(Z-1)^\kappa d\sigma(Z) \sim 0 \tag{7}$$

meaning that all entire functions, integrated against the first member of (7), give 0. Let us write

$$\begin{aligned} Z^z &= \sum_{j=0}^{\infty} \frac{1}{j!}(Z^z)^{(j)}_{Z=1}(Z-1)^j \\ &= 1 + \sum_{j=1}^{\infty} \frac{z(z-1)\cdots(z-j+1)}{j!}(Z-1)^j. \end{aligned}$$

We obtain from (5) and (7)

$$f(z) = R_{\kappa-1}(z) \tag{8}$$

polynomial of degree $\leq \kappa - 1$. Let us state the result.

THEOREM 1. *If (5) holds with $f \in EI$ and the support of $d\sigma$ is inside the open disc $|Z-1| < 1$, f is a polynomial.*

COROLLARY 1.1 *(Pólya's theorem). If $f \in EI$ and $ET(f) < \log 2$, f is a polynomial.*

This comes from the inequality

$$|\log(2\cos\theta) + i\theta| \geq \log 2 \qquad (-\frac{\pi}{2} < \theta < \frac{\pi}{2})$$

(proved on comparing the functions $\log(2\cos\theta)$ and $((\log 2)^2 - \theta^2)^{1/2})$ which expresses that the image of the disc $|\varsigma| < \log 2$ through the map $\varsigma \to e^{\varsigma}$ is contained in the disc $|Z-1| < 1$.

COROLLARY 1.2. *If $f \in EI$, with real exponential type zero and $ET(f) < \frac{\pi}{3}$, f is a polynomial.*

For, again, the image of $CD(f)$ is contained in the disc $|Z-1| < 1$.

The examples 2^z and $2\cos\frac{\pi}{3}z$ show that the inequalities in the corollaries must be strict.

Before considering other specific polynomials $P(Z)$ let us go back to the lemma. Suppose that we are given $P(Z)$ with $|P(Z)| < 1$ on the support of $d\sigma$, and let $N =$ degree of $P(Z)$. Then we have (6) for $\ell = 0, 1, \cdots, N-1$ whenever k is large enough, say $k \geq \kappa$. Therefore the measure $(P(Z))^{\kappa} d\sigma(Z)$ is orthogonal to all polynomials of the form $Z^{\ell}(P(Z))^j$ ($\ell \in \{0, 1, \cdots, N-1\}$, $j \in \mathbb{N}$). Now observe that every polynomial is a linear combination of such polynomials $Z^{\ell}(P(Z))^j$ (obvious by induction on the degree, using division by $P(Z)$). Therefore

$$(P(Z))^{\kappa} d\sigma(Z) \sim 0$$

with the same meaning as in (7).

In $\mathbb{C}\backslash\mathbb{R}^-$ the function $Z \to (P(z))^{-\kappa} Z^z$ is meromorphic and can be written as

$$\sum_{a \in A} a^z Q_a\left(\frac{1}{Z-a}, z\right) + H(Z)$$

where A denotes the zero set of $P(Z)$, $Q_a(\cdot,\cdot)$ is a polynomial in two variables, and $H(Z)$ a holomorphic function in $\mathbb{C}\backslash\mathbb{R}^-$. Integration with respect to $(P(Z))^{\kappa} d\sigma(Z)$ gives

$$f(z) = \sum_{a \in A} a^z R_a(z) \tag{9}$$

where $R_a(z)$ is a polynomial depending on a, provided that $H(Z)$ can be approximated by polynomials uniformly on the support of $d\sigma$. This is the case when the support of $d\sigma$ is contained in $\mathbb{C}\backslash R^-$. For simplicity we make this assumption. (Actually, it suffices that the support of $d\sigma$ does not contain any curve around 0, a condition that $P(0) \neq 0$ implies). Let us state the result.

THEOREM 2. *Suppose that a) $f \in EI$ and (5) holds, the support of $d\sigma$ being contained in $\mathbb{C}\backslash\mathbb{R}^-$; b) $P(Z)$ is a polynomial with integral coefficients such that $|P(Z)| <$ 1 on the support of $d\sigma$. Then $f(z)$ has the form (9), A being the zero set of P.*

Here is a statement in the opposite direction.

THEOREM 2.1. *Let $P(Z) = Z^N + c_1 Z^{N+1} + \cdots + c_N$ be a unitary polynomial with integral coefficients, and $Z(P)$ be the set $\{Z \; ; \; | P(Z) | < 1\}$. Then any open set which contains the closure $\overline{Z(P)}$ contains infinitely many finite sets A such that $f(z) = \sum_{a \in A} a^z$ belongs to EI.*

Proof. The zero sets of $(P(Z))^k + 1$ cluster on the boundary of $Z(P)$ as $k \to \infty$, and for any set of conjugate algebraic integers A, the function $f(z) = \sum_{a \in A} a^z$ belongs to EI.

Theorem 2 takes a more attractive form by using the notion of logarithmic capacity. For compact sets K in $\mathbf{C}$ there are equivalent notions : logarithmic capacity, transfinite diameter, Tchebycheff's constant, mapping radius (when $\mathbf{C} \backslash K$ is a conformal image of the exterior of the unit disc) (see [7]). A way to express that the logarithmic capacity is < 1 is the existence of a unitary polynomial $P(Z)$ such that $| P(Z) | < 1$ on K. Here is an important result of Fekete [2].

FEKETE'S LEMMA (1926). *If the logarithmic capacity of a compact set K is strictly less than 1, there exists a unitary polynomial $P(Z)$ with integral coefficients such that $| P(Z) | < 1$ on K.*

[Fekete states the result with coefficients in $\mathbb{Z}(i)$, but the proof gives also coefficients in $\mathbb{Z}$].

Using Fekete's lemma, theorem 2 provides the main theorem of this theory.

THEOREM 3. *Suppose that a) $f \in EI$ and (5) holds, the support of $d\sigma$ being contained in $\mathbf{C} \backslash \mathbb{R}^-$; b) the support of $d\sigma$ has logarithmic capacity < 1. Then $f(z)$ has the form (9), A being a finite union of sets of conjugate algebraic integers.*

COROLLARY 3.1. *(Theorem of Pisot) Let τ_0 be the positive number such that the logarithmic capacity of the image of the disc $\{\varsigma \; ; \; | \varsigma | \leq \tau_0\}$ by the exponential mapping $\varsigma \to e^\varsigma$ is equal to 1. Given any $\tau < \tau_0$, each $f \in EI$ such that $ET(f) < \tau$ has the form (9), where $A = A(\tau)$ is a finite set of algebraic integers* [3].

COROLLARY 3.2. *(Robinson) Given any $\tau < \pi$, each $f \in EI$ such that the real exponential type is 0 and $ET(f) < \tau$ has the form (9), where $A = A(\tau)$ is a finite set of points $e^{i\pi r}$, r rational* [6].

Actually we used a theorem of Kronecker (1857) which says that algebraic integers carried by the circle $\mid Z \mid = 1$ are of the form $e^{i\pi r}$, plus the fact that the logarithmic capacity of the unit circle is 1.

Corollaries 3.1 and 3.2 express that there is a discrete scale of possible types for the functions $f \in EI$ (resp. for the functions $f \in EI$ with vanishing real exponential type).

Pisot gave the estimate $\tau_0 = 0,843\cdots$ [4]. However only the first steps of the scale are known :

$$ET(1) = 0$$
$$ET(2^z) = \log 2 = 0,693$$
$$ET(\gamma^z + \bar{\gamma}^z) = 0,759\cdots \quad \text{where } \gamma = 1 + e^{i\frac{\pi}{3}}.$$

Every $f \in EI$ such that $ET(f) = 0,82$ can be written as

$$R_0(z) + 2^z R_1(z) + \gamma^z R_2(z) + \bar{\gamma}^z R_3(z)$$

where the $R_j(z)$ are polynomials [3] (Pisot assumes $ET(f) < 0,8$ and the slight improvement comes from complicated polynomials $P(z)$). Are there possible types between 0,82 and τ_0 ? According to [6] this is an open question. On the other hand, Robinson observed that there are infinitely many types between τ_0 and $\tau_0 + \epsilon$ [6]. This will appear as a consequence of theorem 4 below.

For the functions $f \in EI$ with vanishing real exponential type the situation is much simpler : the scale is perfectly known :

$$ET(1) = 0$$
$$ET(2\cos\frac{\pi}{3}z) = \frac{\pi}{3} = \pi - \frac{2\pi}{3}$$
$$ET(\cos\frac{\pi}{2}z) = \frac{\pi}{2} = \pi - \frac{2\pi}{4}$$
$$ET(\cos\frac{\pi}{10}z + \cos\frac{3\pi}{10}z) = \pi - \frac{2\pi}{5}$$
$$ET(2\cos\frac{2\pi}{3}z) = \pi - \frac{2\pi}{6}$$

and so on. Let us observe that $\frac{\sin \pi z}{\pi z}$ belongs to EI, as well as $e^{\alpha z}\frac{\sin \pi z}{\pi z}$ for any α. It is a case when condition a) in theorem 2 (or the substitutes mentioned before the theorem) do not apply.

THEOREM 4. *Let K be a compact set in C, of logarithmic capacity equal to 1. Given an open set $G \supset K$, there exist infinitely many sets A contained in G such that $f(z) = \sum_{a \in A} a^z$ belongs to EI.*

Proof. First we take $K_0 \subset K$, so that its logarithmic capacity lies between $1-\epsilon$ and 1. Applying Fekete's lemma we obtain a unitary polynomial $P(Z)$ with integral coefficients such that $K_0 \subset Z(P)$. If ϵ is small enough this implies $Z(P) \subset G$, and we conclude as in theorem 2.1.

References

[1] R. B. BOAS *Entire functions.* Academic Press 1954.

[2] M. FEKETE *Potenzreihen mit ganzzahligen Koefficienten.* Math. Annalen 96 (1926), 410-417.

[3] C. PISOT *Ueber ganzwertige ganze Funktionen.* Jahresbericht der deutchen math. Vereinigung 52 (1942), 95-102.

[4] C. PISOT *Sur les fonctions arithmétiques à croissance exponentielle.* C. R. Acad. Sc. Paris 222 (1946), 988-990.

[5] G. PÓLYA *Collected works, volume I.*

[6] R. M. ROBINSON *Integer-valued entire functions.* Trans. Amer. Math. Soc. 153 (1971), 451-468.

[7] M. TSUJI *Potential theory in modern function theory.* Tokyo 1959.

CNRS URA D0757
Université de Paris-Sud
Mathématiques - Bât. 425
91405 ORSAY CEDEX (France)

ALGEBRA DIRECT SUM DECOMPOSITION OF $C_R(X)$

M.H. VASAVADA* AND R.D. MEHTA

DEDICATED TO THE MEMORY OF U.N. SINGH

INTRODUCTION Let X be a compact Hausdorff space and let $C_R(X)$ denote the Banach algebra of all real-valued continous functions on X with supremum norm. Throughout this paper A and B denote closed subalgebras of $C_R(X)$ with $1 \in A$ and $1 \notin B$. Let Z(B) denote the set of common zeros of functions of B. Since $1 \notin B$, $Z(B) \neq \phi$ [2, Lemma 1.1]. If $A \oplus B = C_R(X)$ and if B is a closed ideal, then Z(B) is a retract of X [4]. What can be said when B is closed algebra, not necessarily an ideal ? This question has been studied in [1] and [3] and the first section of this paper has been devoted to the further study of this question. Also, if K is a closed subset of X which is a retract of X, then there exists a closed algebra A with unit and a closed ideal B such that $A \oplus B = C_R(X)$ and Z(B) = K. In fact, we can take $B = \{ g \in C_R(X): g(x) = o$ for all $x \in K \}$ and $A = \{ h \circ \sigma : h \in C_R(X) \}$ where σ is a retraction of X onto K. Thus a closed subset K of X and a map of X onto K determine a pair of closed algebra and a closed ideal whose direct sum is $C_R(X)$. What kind of closed sets and maps

---

1980 Mathematics Subject Classification (1985 Revision)
Primary 46J10.

* Research supported by University Grants Commission of India grant F-26-1(2208)86(SR IV).

related to these sets determine a pair of closed subalgebras whose direct sum is $C_R(X)$? The second section of this paper deals with this question.

We need some definitions and notations. For x, y $\in$ X, we say that x is A-related to y, and write $x \overset{A}{\text{—}} y$, if f(x) = f(y) for every f in A. 'B-related' is defined similarly. For $x \neq y$, we say that there is an (A,B) chain of length 1 from x to y if $x \overset{A}{\text{—}} y$ or $x \overset{B}{\text{—}} y$ and that there is an (A,B) chain of length $n (n \geq 2)$ from x to y if there is an ordered set of distinct points $x = x_o, x_1, x_2, \ldots, x_n = y$ with $x_i \overset{A}{\text{—}} x_{i+1} \overset{B}{\text{—}} x_{i+2}$ or $x_i \overset{B}{\text{—}} x_{i+1} \overset{A}{\text{—}} x_{i+2}$ (i = 0, 1, 2,...,n - 2). Fisher [1] proved that when $A \oplus B = C_R(X)$, there exists an (A,B) chain from each point not in Z(B) to a point in Z(B) of length less than or equal to $||P_A||$ where P_A is the projection of $C_R(X)$ onto A. It was shown in [2] that such a chain is unique. If a chain from a point x not in Z(B) ends at a point y in Z(B), we can define $\tau(x) = y$. By taking $\tau(x) = x$ for $x \in zC_P$, we have, when $A \oplus B = C_R(X)$, a map τ from X onto Z(B) which is identity on Z(B). One would hope that this map τ is continuous on X so that Z(B) is a retract of X, as is the case when B is a closed ideal. This hope is falsified by the following example.

EXAMPLE 1. Let X = [-1, 1], A = { $f \in C_R$ [-1, 1]: f is constant on [0, 1] and B = { $g \in C_R$[-1, 1]: g is even and g(1) = o }. Then A and B are closed subalgebras of $C_R(X)$, $1 \in A$, $1 \notin B$ and $A \oplus B = C_R(X)$. It can be easily seen that $\tau(x) = 1$ if $-1 < x \leq 1$ while $\tau(-1) = -1$. Hence τ is not continuous on [-1, 1]. Clearly Z(B) = {-1, 1} is not a retract of X.

In the above example, τ is discontinuous at -1 which is a point in Z(B). We shall show that this situation is typical in the sense that if τ is discontinuous on X, it has to be discontinuous at some point of Z(B).

We need a few more definitions and notations. Assume that $A \oplus B = C_R(X)$ and that the maximum of lengths of (A,B) chains from points not in Z(B) to points in Z(B) is n. Let $S_{-1} = \phi$, $S_o = Z(B)$ and define S_i ($i = 1, 2, \ldots, n$) inductively by $S_i = \{ x \in X: x \overset{C}{—} y$ for some $y \in S_{i-1}\}$ where C = A or B according as i is odd or even. By [3, Lemma 3] each S_i is a closed set. Also $S_{-1} \subset S_o \subset \ldots \subset S_n = X$. We note that given $x \in S_i - S_{i-1}$ we can get only one $y \in S_{i-1}$ such that $x \overset{C}{—} y$ (C = A or B) by the uniqueness of the chain. For $i = 1, 2, \ldots, n$ define the maps σ_i of S_i onto S_{i-1} by

$$\sigma_i(x) = y \text{ if } x \in S_i - S_{i-1} \text{ and } x \overset{A}{——} y \text{ or } x \overset{B}{——} y$$
$$= x \text{ if } x \in S_{i-1}.$$

By the above remarks, σ_i's are well defined. Finally we note that the map τ defined earlier, from X onto Z(B), is given by $\tau = \sigma_1 \circ \sigma_2 \circ \ldots \sigma_n$ and if $x \in S_i$ ($1 \le i \le n$), then $\tau(x) = (\sigma_1 \circ \sigma_2 \circ \ldots \circ \sigma_i)(x)$.

1. CONTINUITY OF τ

THEOREM 2. Let $A \oplus B = C_R(X)$.

(a) For $x \in X$, τ is continuous at x if τ is continuous at $\tau(x)$.

(b) If τ is continuous on Z(B), then Z(B) is a retract of X.

PROOF Suppose that $x \in X$ and τ is continuous at $\tau(x)$.
If $x \in S_o$, then $\tau(x) = x$ and hence τ is continuous at x.
So let $x \in S_j - S_{j-1}$ where $j \ge 1$. To show that τ is continuous at x, it is enough to prove that if (x_α) is a net in $S_i - S_{i-1}$($j \le i \le n$) converging to x, then the net $(\tau(x_\alpha))$ converges to $(\tau(x))$. For such a net (x_α), let $y_\alpha = \sigma_i(x_\alpha)$, then $\tau(x_\alpha) = (\sigma_1 \circ \sigma_2 \circ \ldots \circ \sigma_i)(x_\alpha) = (\sigma_1 \circ \sigma_2 \circ \ldots \sigma_{i-1})(y_\alpha) = \tau(y_\alpha)$ if $i \ge 2$ and $\tau(x_\alpha) = \sigma(x_\alpha)(x_\alpha) = y_\alpha = \tau(y_\alpha)$ if $i = 1$. Now suppose that (y_α) converges to y. Then $x_\alpha \overset{A}{——} y_\alpha$ implies that $x \overset{A}{——} y$ and $x_\alpha \overset{B}{——} y_\alpha$

implies that $x \overset{B}{\text{——}} y$. If $y \in S_j - S_{j-1}$ and $y \neq x$, then we have two different chains from x to $\tau(x)$, which is not possible. Hence we must have $y = x$ or $x = \sigma_{j+1}(y)$ or $y = \sigma_j(x)$. This gives $\tau(y) = \tau(x)$.

Now let $x_\alpha = x_\alpha^o$, $\sigma_i(x_\alpha^o) = x_\alpha^1$, $\sigma_{i-1}(x_\alpha^1) = x_\alpha^2$, ... , $\sigma_1(x_\alpha^{i-1}) = x_\alpha^i$, and suppose that the net (x_α^k) converges to x ($1 \leq k \leq i$). Also let $x_o = x$. Then $x \notin S_o$ and $x_i \in S_o$. Let m be the least positive integer such that $x_m \in S_o$. Then by what we have seen above,

(1) $$\tau(x_\alpha) = \tau(x_\alpha^o) = \tau(x_\alpha^1) = \ldots = \tau(x_\alpha^m)$$

and

(2) $$\tau(x) = \tau(x_o) = \tau(x_1) = \ldots = \tau(x_m)$$

Since $x_m \in S_o$, by (2) $\tau(x) = \tau(x_m) = x_m$ and so τ is continuous at x Hence by (1) and (2),

$$\tau(x) = \tau(x_m) = \operatorname*{Lim}_\alpha \tau(x_\alpha^m) = \lim_\alpha \tau(x_\alpha).$$

This completes the proof of (a).

If now τ is continuous on Z(B), it is continuous on X by (a). Since $\tau(x) = x$ for each x in Z(B), τ is a retraction of X onto Z(B), which proves (b).

EXAMPLE 3. Let $E_1 = \{(x,o) \in \mathbb{R}^2 : o \leq x \leq 1\}$,

$E = \{(x,o) \in \mathbb{R}^2 : -1 \leq x \leq o\}$,

$E = \{(-1,y) \in \mathbb{R}^2 : o \leq y \leq 1\}$ and $X = E_1 \cup E_2 \cup E_3$.

Let $A = \{ f \in C_R(X)$: f is constant on E_1 and f is constant on $E_3 \}$ and $B = \{ g \in C_R(X)$: $g((x,o)) = g(-x, o))$ for $o \leq x \leq 1$ and $g(1,0) = 0 \}$.

Then $A \oplus B = C_R(X)$. If $p = (-1,1/2)$, then τ is continuous at p, but not continuous at $\tau(p) = (-1,0)$.

The above examples shows that the converse of part (a) of the Theorem 2 does not hold.

2. CONDITIONS FOR ALGEBRA DIRECT SUM DECOMPOSITION OF $C_R(X)$.

THEOREM 4. Let $S_o, S_1, \ldots, S_n$ be nonempty closed subset of X such that $\phi = S_{-1} \subset S_o \subset S_1 \subset \ldots \subset S_n = X$ and let

$\sigma_i : S_i \to S_{i-1}$ (i = 1, 2, ..., n) be maps such that

$\sigma_i(x) \in S_{i-1} - S_{i-2}$ if $x \in S_i - S_{i-1}$ and $\sigma_i(x) = x$ if $x \in S_{i-1}$.

Suppose, in addition that the sets $\{S_i\}_{i=1}^n$ and the maps $\{\sigma_{i=1}^n\}$ satisfy the following condition:

If a net (x_α) in $S_i - S_{i-1} (1 \leq i \leq n)$ converges to $x \in S_j - S_{j-1}$ $(o \leq j \leq i)$ and if the net $(y_\alpha = \sigma_i(x_\alpha))$ converges to $y \neq x$, then

(i) $y \in S_{j-1} - S_{j-2}$ and $\sigma_j(y) = \sigma_j(x)$ whenever $1 \leq j \leq i$ and $i + j$ is even;

(ii) $y \in S_{j+1} - S_j$ and $\sigma_{j+1}(y) = \sigma_{j+1}(x)$ whenever $o \leq j \leq i$ and $i+j$ is odd;

(iii) $y \in S_o$ if $j = o$ and i is even.

Then there exist closed subalgebras A, B of $C_R(X)$ such that $1 \in A$, $1 \notin B$, $A \oplus B = C_R(X)$ and $Z(B) = S_o$.

PROOF Let $A_o = C_R(S_o)$, $B_o = \{0\}$. Then $1 \in A_o$, $1 \notin B_o$, A_o and B_o are closed subalgebras of $C_R(S_o)$, $A_o \oplus B_o = C_R(S_o)$ and $Z(B_o) = S_o$. Suppose that for some k, $o \leq k \leq n - 1$, the closed subalgebras A_i, B_i of $C_R(S_i)$ have been defined for i = 0, 1, 2, ..., k so that $1 \in A_i$

, $1 \in B_i$, $A_i + B_i = C_R(S_i)$ and $Z(B_i) = S_o$. Then define the sets A_{k+1}, B_{k+1} of real-valued functions on S_{k+1} as follows:

For k even,

$$A_{k+1} = \{ f \circ \sigma_{k+1} : f \in A_k \}, \quad B_{k+1} = \{ g \in C_R(S_{k+1}) : g|S_k \in B_k \}$$

and for k odd,

$$A_{k+1} = \{ f \in C_R(S_{k+1}) : f|S_k \in A_k \}, \quad B_{k+1} = \{ g \circ \sigma_{k+1} : g \in B_k \}$$

Let $\phi \in A$, $\psi \in B_{k+1}$. Then it is clear that $\phi|S_j \in A_j$, $\psi|S_j \in B$ for $o \leq j \leq k + 1$ and hence

$$\phi(x) = \phi(\sigma_j(x)) \text{ for } x \in S_j - S_{j-1},\ j \text{ odd},\ 1 \leq j \leq k+1,$$

$$\psi(x) = \psi(\sigma_j(x)) \text{ for } x \in S_j - S_{j-1},\ j \text{ even},\ 2 \leq j \leq k+1,$$

$$\psi(x) = o \text{ for } x \in S_o$$

We show that $A_{k+1} \subset C_R(S_{k+1})$. If k is odd, this is obvious. So assume that k is even. Let $\phi \in A_{k+1}$. Then there exists $f \in A_k$ such that $\phi = f \circ \sigma_{k+1}$. Let $x \in S_{k+1}$ and (x_α) be a net in S_{k+1} converging to x. If $(x_\alpha) \subset S_k$, then we have $\phi(x_\alpha) =. f(\sigma_{k+1}(x_\alpha)) = f(x_\alpha) \to f(x) = f(\sigma_{k+1}(x)) = \phi(x)$. Suppose that $(x_\alpha) \subset S_{k+1} - S_k$. Let $y_\alpha = \sigma_{k+1}(x_\alpha)$ and suppose that $y_\alpha \to S$. Then $y_\alpha \in S_k - S_{k-1}$ and $y \in S_k$. Let $x \in s_j - S_{j-1}$. First assume that j is even, $o \leq j \leq k$. Then $k + 1 + j$ is odd and hence by hypothesis $y = x$ or $y \in S_{j+1} - S_j$ and $\sigma_{j+1}(y) = \sigma_{j+1}(x)$. Since $y \in S_k$, we have $j + 1 \leq k$. Hence by (*), $\phi(x_\alpha) = f(\sigma_{k+1}(x_\alpha)) = \phi(x)$. Next let j be odd ($1 \leq j \leq k+1$). Then $k + 1 + j$ is even, so that $y = x$ or $y \in S_{j-1} - S_{j-2}$ and $\sigma_j(y) = \sigma_j(x)$. Hence again by (*) $\phi(x_\sigma) = f(\sigma_{k+1}(x_\alpha)) = f(y_\alpha) \to f(y) = f(\sigma(y)) = \phi(y) = \phi(\sigma_j(y)) = \phi(\sigma_j(x) = \phi(x)$. This shows that ϕ is continuous at x and hence on S_{k+1}. Thus $A_{k+1} \subset C_R(S_{k+1})$.

Similarly it can be shown that $B_{k+1} \subset C_R(S_{k+1})$. It is easy to see that A_{k+1}, B_{k+1} are closed subalgebras of $C_R(S_{k+1})$, $1 \in A_{k+1}$, $A_{k+1} \cap B_{k+1} = \{ 0 \}$, $A_{k+1} \oplus B_{k+1} = C_R(S_{k+1})$ and $Z(B_{k+1}) = S_o$. By induction we shall have A_n, B_n and taking $A = A_n$ and $B = B_n$ we get the required result.

REMARK 5. We have defined the sets S_i and maps σ_i in the introduction, when $A + B = C_R(X)$. It can be verified that these sets and maps satisfy the condition of the Theorem 4. Thus we have given necessary and sufficient conditions for algebra direct sum decomposition of $C_R(X)$ in terms of a finite increasing sequence of closed subsets of X and maps associated with them.

REFERENCES

1. S.D. Fisher, The decomposition of $C_{\mathbb{R}}(K)$ into the direct sum of subalgebras, J.Funct. Anal. 31(1979), 218 - 223.

2. R.D. Mehta and M.H. Vasavada, Algebra direct sum decomposition of $C_R(X)$, Proc. Amer. Math. Soc. 98(1986) 71 - 74.

3. ____________ , Algebra direct sum decomposition of $C_{\mathbb{R}}(X)$. II, Proc. Amer. Math. SOC. 100 (1987), 123 - 126.

4. H. Yoshizawa, On simultaneous extension of continuous functions, Proc. Imp. Acad. Tokya 20 (1944), 653 - 654.

DEPARTMENT OF MATHEMATICS,
SARDAR PATEL UNIVERSITY,
VALLABH VIDYANAGAR - 388 120,
GUJRAT, INDIA.

Exposed Points and Points of Continuity in Closed Bounded Convex Sets

Pradipta Bandyopadhyaya
(Dedicated to the memory of Prof. U. N. Singh)

Introduction and Preliminaries

To fix our notations and conventions first, we work only with *real* Banach spaces. For a Banach space X and $A \subseteq X$, we denote (i) by $\text{co}(A)$ the convex hull of A, (ii) by $\text{sp}(A)$ the linear span of A, (iii) by A^o the *polar* of A, i.e., $A^o = \{f \in X^* : f(x) \leq 1 \text{ for all } x \in A\}$, and (iv) by $\overline{A}^\tau$ the closure of A for the topology τ on X. Whenever the topology is not specified, we mean the *norm* topology. For $z \in X$ and $r > 0$, we denote by $B_r[z]$ (resp. $B_r(z)$) the closed (resp. open) ball of radius r and centre z.

If K is a closed bounded convex set in X and $f \in X^*$, let $M(K, f) = \sup\{f(x) : x \in K\}$. If $\alpha > 0$, a *slice* of K determined by f and α is $S(K, f, \alpha) = \{x \in K : f(x) > M(K, f) - \alpha\}$. $S(K, f, \alpha)$ is a non-empty, (relatively) weak open subset of K. Recall (from [2]) that $x_o \in K$ is called

(a) a *denting point of* K if, for each $\varepsilon > 0$, $x_o \notin \overline{\text{co}}(K \setminus B_\varepsilon(x_o))$.

(b) a *τ-point of continuity (τ-PC) for* K if the identity map, $id : (K, \tau) \longrightarrow (K, \text{norm})$ is continuous at x_o. A weak-PC is simply called a PC.

(c) an *exposed point of* K if there exists $f \in X^*$ such that f exposes x_o, i.e., $f(x) < f(x_o)$ for all $x \in K \setminus \{x_o\}$.

AMS(MOS) subject classifications (1980) : 46B20

Keywords and Phrases : Exposed Points, Strongly Exposed Points, Denting Points, Points of Continuity, Banach Spaces Containing ℓ^1

(d) a *strongly exposed point of* K if there exists $f \in X^*$ such that f exposes x_o and for any sequence $\{x_n\} \subseteq K$, $\lim_n f(x_n) = f(x_o)$ implies that $\lim_n \|x_n - x_o\| = 0$.

When X is a dual space and in (c) (resp. (d)) f comes from the predual of X, we say x_o is w*-exposed (resp. w*-strongly exposed).

We will say that $f \in X^*$ *supports* K *at* x_o if $f(x_o) = M(K, f)$ and we let $S(K, x_o) = \{f \in X^* : f(x_o) = M(K, f)\}$. Note that $S(K, x_o)$ is a w*-closed convex cone with vertex 0. In the sequel, we may sometimes abbreviate $S(K, x_o)$ by $S(x_o)$ when there is no scope of confusion.

Call $x_o \in K$ a *relatively exposed point of* K if for all $x \in K \setminus \{x_o\}$ there exists $f \in S(K, x_o)$ such that $f(x) < f(x_o)$. Call x_o a *vertex of* K if for all $x \in X \setminus \{x_o\}$ there exists $f \in S(K, x_o)$ such that $f(x) \neq f(x_o)$, or, in other words, $S(K, x_o)$ separates points of X. Again, we can define w*-relatively exposed points or w*-vertices similarly. Obviously, a (w*-) vertex is (w*-) relatively exposed.

Let $\widehat{K}$ be the image of $K \subseteq X$ under the canonical embedding of X in X^{**}. We denote by $\widetilde{K}$ the w*-closure of $\widehat{K}$ in X^{**}. $\widetilde{K}$ is of course a w*-compact convex set.

It is known that $x_o \in K$ is denting if and only if x_o is contained in slices of K of *arbitrarily small* (norm-) diameter (see e.g., [8]). Similarly, one can show that $x_o \in K$ is strongly exposed by $f \in X^*$ if and only if x_o is contained in slices of K *determined by* f of arbitrarily small diameter, or, equivalently, $\lim_{\alpha \to 0^+} \operatorname{diam}[S(K, f, \alpha)] = 0$. From these observations, the following implications are easy to establish :

$$\begin{array}{ccc} x_o \text{ strongly exposed} & \Longrightarrow & x_o \text{ denting point} \\ \Downarrow & & \Downarrow \\ x_o \text{ exposed and PC} & \Longrightarrow & x_o \text{ extreme and PC} \end{array}$$

A recent result (see [7] and [8]) is that x_o is denting if and only if x_o is extreme and PC, i.e., the implication down the right hand side above is reversible. This naturally leads to the conjecture that the implication down the left hand side above, too, can be reversed, i.e., if x_o is exposed and PC (or, equivalently, x_o is an exposed denting point) then x_o is strongly exposed.

We show that the conjecture is false by constructing a counterexample in the Banach space ℓ^1 and provide a characterisation of strongly exposed points among points of continuity of a closed bounded convex set. As a corollary, we deduce that the conjecture is true for weakly compact sets. We also show that the counterexample, in some sense, is actually typical of ℓ^1, i.e., we characterise Banach spaces containing ℓ^1 in terms of the validity of the above conjecture.

1 The Counterexample

For our X, we take ℓ^1. Let $\{\delta_n : n \geq 1\}$ denote the canonical basis of ℓ^1. Recall that ℓ^1 as a dual space has a w*-topology induced on it. Let

$$K = \overline{\text{co}}^{w^*}[\{\frac{1}{n}\delta_1 + \delta_n : n \geq 2\} \cup \{\frac{1}{n^2}\delta_1 - \frac{1}{n}\delta_n : n \geq 2\}]$$

From Milman's Theorem (see [9, p 9]) and the metrizable version of Choquet's Theorem ([1] or [9]), it follows that $x \in K$ if and only if x is of the form

$$x = \sum_{n\geq 2} \alpha_n(\frac{1}{n}\delta_1 + \delta_n) + \sum_{n\geq 2} \beta_n(\frac{1}{n^2}\delta_1 - \frac{1}{n}\delta_n)$$

with $\alpha_n, \beta_n \geq 0$ and $\sum_{n\geq 2}(\alpha_n + \beta_n) \leq 1$ (see [2, Example 3.2.5] for the details in a similar situation). Observe that $(-1, 0, 0, \ldots) \in c_o \subseteq \ell^\infty = (\ell^1)^*$ exposes $0 = (0, 0, \ldots) \in K$. If $f = (a_n) \in \ell^\infty$ supports K at 0, then $\frac{1}{n}a_1 \pm a_n \leq 0$ for $n \geq 2$, from which it follows that $a_1 \leq 0$ and $|a_n| \leq \frac{|a_1|}{n}$, $(n \geq 2)$ and we see therefore that $(a_n) \in c_o$. Since $f(\frac{1}{m}\delta_1 + \delta_m) = \frac{a_1}{m} + a_m \to 0$ as $m \to \infty$, $S(K, f, \alpha) = \{y \in K : f(y) > -\alpha\}$ contains $\frac{1}{m}\delta_1 + \delta_m$ for all sufficiently large m, whence $\text{diam}[S(K, f, \alpha)] \geq \|\frac{1}{m}\delta_1 + \delta_m\| > 1$, showing that 0 is not strongly exposed.

Next, for $m \geq 2$, let $f_m = (-1, -\frac{1}{m^2}, \ldots, -\frac{1}{m^2}, -\frac{1}{m}, -\frac{1}{m}, \ldots) \in \ell^\infty$ where the mth coordinate onwards of f_m has the value $-\frac{1}{m}$ and consider the slice $S_m = \{y \in K : f_m(y) > -\frac{1}{m^3}\}$ determined by f_m. Clearly, $0 \in S_m$ and if

$$x = \sum_{n\geq 2} \alpha_n(\frac{1}{n}\delta_1 + \delta_n) + \sum_{n\geq 2} \beta_n(\frac{1}{n^2}\delta_1 - \frac{1}{n}\delta_n) \in S_m$$

we must have

$$-\frac{1}{m^3} < -\sum_{n=2}^{m-1} \alpha_n \cdot \frac{m^2+n}{m^2 n} - \sum_{n\geq m} \alpha_n \cdot \frac{m+n}{mn} - \sum_{n=2}^{m-1} \beta_n \cdot \frac{m^2-n}{m^2 n^2} + \sum_{n\geq m} \beta_n \cdot \frac{n-m}{n^2 m}$$

whence it follows that

$$\begin{aligned}
\frac{1}{m^2} \;&>\; \sum_{n=2}^{m-1} \alpha_n \cdot \frac{n+1}{n} \cdot \frac{m^2+n}{m(n+1)} + \sum_{n\geq m} \alpha_n \cdot \frac{n+1}{n} \cdot \frac{m+n}{n+1} \\
&\quad + \sum_{n=2}^{m-1} \beta_n \cdot \frac{n+1}{n^2} \cdot \frac{m^2-n}{m(n+1)} - \sum_{n\geq m} \beta_n \cdot \frac{n-m}{n^2} \\
&\geq\; \sum_{n=2}^{m-1} \alpha_n \cdot \frac{n+1}{n} + \sum_{n\geq m} \alpha_n \cdot \frac{n+1}{n} + \frac{m-1}{m+1} \sum_{n=2}^{m-1} \beta_n \cdot \frac{n+1}{n^2} - \frac{1}{4m}
\end{aligned}$$

on using successively the following easily verified inequalities :

(i) for all $m \geq 1$ and $n \leq m$, $\dfrac{m^2+n}{m(n+1)} \geq 1$.

(ii) for all $m \geq 1$, $\dfrac{m+n}{n+1} \geq 1$.

(iii) for all $n \leq m$, $\dfrac{m^2-n}{m(n+1)} \geq \dfrac{m-1}{m+1}$, since $\dfrac{m^2-n}{m(n+1)}$ is decreasing in n.

(iv) for all $m, n \geq 1$, $\dfrac{n-m}{n^2} \leq \dfrac{1}{4m}$.

Thus,

$$\begin{aligned} \frac{m+1}{m-1} \cdot \frac{m+4}{4m^2} &\geq \frac{m+1}{m-1} \sum_{n\geq 2} \alpha_n \cdot \frac{n+1}{n} + \sum_{n=2}^{m-1} \beta_n \cdot \frac{n+1}{n^2} \\ &\geq \sum_{n\geq 2} \alpha_n \cdot \frac{n+1}{n} + \sum_{n=2}^{m-1} \beta_n \cdot \frac{n+1}{n^2}. \end{aligned}$$

Now,

$$\begin{aligned} \|x\| &= \sum_{n\geq 2} \left(\frac{\alpha_n}{n} + \frac{\beta_n}{n^2} \right) + \sum_{n\geq 2} \left| \alpha_n - \frac{\beta_n}{n} \right| \leq \sum_{n\geq 2} \alpha_n \cdot \frac{n+1}{n} + \sum_{n\geq 2} \beta_n \cdot \frac{n+1}{n^2} \\ &\leq \frac{(m+1)(m+4)}{4m^2(m-1)} + \sum_{n\geq m} \beta_n \cdot \frac{n+1}{n^2} \leq \frac{(m+1)(m+4)}{4m^2(m-1)} + \frac{m+1}{m^2} \\ &= \frac{5(m+1)}{4m(m-1)} \leq \frac{15}{4m} \quad \text{for all } m \geq 2. \end{aligned}$$

Hence, $\operatorname{diam}(S_m) \leq \dfrac{15}{2m} \longrightarrow 0$ as $m \longrightarrow \infty$ and we conclude that 0 is a denting point of K.

REMARKS : 1. Let F_o be a w*-cluster point of $\{\frac{1}{n}\delta_1 + \delta_n : n \geq 2\}$ in $\widetilde{K}$. Then $F_o \neq 0$ since $F_o(f_o) = 1$ where $f_o = (1,1,\ldots,1,\ldots) \in \ell^\infty$. As $S(0) \subseteq c_o$, we must have $F_o \in \cap\{\ker(\hat{f}) : f \in S(0)\}$ and hence 0 is *not* w*-relatively exposed in $\widetilde{K}$. This observation proved useful in the formulation (c) of Theorem 1 in the next section.

2. Since $S(0) \subseteq c_o$, one in fact has $S(0) = \{(a_n) \in c_o : a_1 \leq 0$ and $|a_n| \leq \frac{|a_1|}{n}, n \geq 2\}$. It is now easy to see that $S(0)$ separates points of ℓ^1, i.e., 0 is a vertex of K.

2 A Characterisation Theorem

In this section, we prove the following theorem :

Theorem 1 *Let K be a closed bounded convex set in a real Banach space X and $x_o \in K$ be a PC. The following are equivalent :*

(*a*) *$x_o \in K$ is strongly exposed.*

(*b*) *$\hat{x}_o \in \widetilde{K}$ is w^*-exposed.*

(*c*) *$\hat{x}_o \in \widetilde{K}$ is w^*-relatively exposed.*

(*d*) *$X^*/\overline{\mathrm{sp}}(S(x_o)) = \overline{\pi[(K-x_o)^o]}$ where $\pi : X^* \longrightarrow X^*/\overline{\mathrm{sp}}(S(x_o))$ is the quotient map.*

We collect in the following lemma some facts that will be used in the proof of Theorem 1.

Lemma 2 *Let K be a closed bounded convex set in a real Banach space X. Then*

(*a*) *$F_o \in \widetilde{K}$ is a w^*-PC if and only if there exists $x_o \in K$ such that $\hat{x}_o = F_o$ and x_o is a PC.*

(*b*) *$\hat{x}_o \in \widetilde{K}$ is w^*-exposed by $f \in X^*$ if and only if $\{S(K,f,\alpha) : \alpha > 0\}$ forms a local base for the relative weak topology on K at x_o.*

(*c*) *A w^*-relatively exposed point $F_o \in \widetilde{K}$ is w^*-exposed if and only if F_o is a w^*-G_δ point.*

Proof : Recall that $\widehat{K}$ is w*-dense in $\widetilde{K}$ and the w*-topology on $\widehat{K}$ coincides with the weak topology on K under the canonical homeomorphism.

Now (a) follows immediately from the relevant definitions, while (b) uses the additional fact that $\widetilde{K}$ is w*-compact.

For (c), observe that the proof given in [1, p 119] for exposed and relatively exposed points defined with respect to continuous affine functions on a compact convex set can be easily adapted to our present context where we deal only with w*-continuous linear functionals on a w*-compact convex set. We omit the details. ■

Now we are in a position to prove Theorem 1.

Proof of Theorem 1 : (*a*) $\Longleftrightarrow$ (*b*) $\Longleftrightarrow$ (*c*) follows from Lemma 2 and relevant definitions.

(*c*) $\Longleftrightarrow$ (*d*) : Using definitions, simple separation arguments and polar calculations *à la* the Bipolar Theorem, each of the statement below is easily seen to be equivalent to the next :

(1) $\hat{x}_o$ is w*-relatively exposed in $\widetilde{K}$.

(2) 0 is w*-relatively exposed in $\widetilde{K}_1$, where $K_1 = K - x_o$.

(3) $\widetilde{K}_1 \cap [\cap\{\ker(\hat{f}) : f \in S(x_o)\}] = \{0\}$.

(4) $K_1^{oo} \cap [\overline{\mathrm{sp}}(S(x_o))]^o = \{0\}$, $\cdots(*)$
(since $\cap\{\ker(\hat{f}) : f \in S(x_o)\} = [\overline{\mathrm{sp}}(S(x_o))]^{\perp} = [\overline{\mathrm{sp}}(S(x_o))]^o$ and $\widetilde{K}_1 = K_1^{oo}$.)

(5) $[K_1^o \cup \overline{\mathrm{sp}}(S(x_o))]^o = \{0\}$.

(6) $\overline{\mathrm{co}}[K_1^o \cup \overline{\mathrm{sp}}(S(x_o))] = X^*$.

This last condition implies that $\overline{[\overline{\mathrm{sp}}(S(x_o)) + K_1^o]} = X^*$ which in turn means that

$$X^*/\overline{\mathrm{sp}}(S(x_o)) = \overline{\pi(K_1^o)}. \qquad \cdots(**)$$

Conversely, suppose that $(**)$ holds and that $\phi \in K_1^{oo} \cap [\overline{\mathrm{sp}}(S(x_o))]^o$. We get immediately

(a) $\phi \in [\overline{\mathrm{sp}}(S(x_o))]^o = [X^*/\overline{\mathrm{sp}}(S(x_o))]^*$, and

(b) $M(K_1^o, \phi) \leq 1$.

But (a) and (b) together give $M(\overline{\pi(K_1^o)}, \phi) \leq 1$ and using $(**)$, we see that $\phi = 0$, thus $(*)$ holds.
This completes the proof of the theorem. ∎

Corollary 3 *If K is weakly compact then $x_o \in K$ is strongly exposed if and only if x_o is exposed and PC.*

Proof : We simply observe that if K is weakly compact, $\widetilde{K} = \widehat{K}$ and that $\hat{x}_o \in \widetilde{K}$ is w*-exposed if $x_o \in K$ is exposed. Now, the corollary follows immediately from Theorem 1. ∎

REMARKS : 1. Note that the proof of (c) $\Longleftrightarrow$ (d) does not use the fact that x_o is a PC.

2. If x_o is a PC and $\hat{x}_o \in \widetilde{K}$ is a w*-vertex, or equivalently, x_o is a PC and $\mathrm{sp}(S(x_o))$ is dense in X^*, it follows from Theorem 1 (c), or (d), that x_o is strongly exposed. But, if $x_o \in K$ is a strongly exposed point with a unique (upto scalar multiples) exposing functional — this happens for instance for any boundary point of the unit ball of the Euclidean space $\mathbb{R}^2$ — then $\mathrm{sp}(S(x_o))$ is 1-dimensional and thus $X^* \neq \overline{\mathrm{sp}}(S(x_o))$, i.e., this condition is, in general, not necessary. And since 0 in our example is a vertex (see Remark 2), the weaker condition that x_o is a PC and $x_o \in K$ is a vertex is no longer sufficient for x_o to be strongly exposed.

3. As we have mentioned already $x_o \in K$ is denting if and only if x_o is contained in slices of K of arbitrarily small diameter. We see from Theorem 1 that $x_o \in K$ is strongly exposed if and only if x_o is contained in slices of K determined by functionals from $S(K, x_o)$ of arbitrarily small diameter. This fact, however, has a direct elementary proof.

3 A Geometric Characterisation of Banach Spaces Containing ℓ^1

Let us say that a closed bounded convex set K in a Banach space X has Property (P) if every exposed PC in K is strongly exposed; and that a Banach space X has Property (P) if every closed bounded convex set K in X has Property (P).

It follows from the counterexample and Corollary 3 above that

(i) If X has the Property (P) then X does not contain a copy of ℓ^1.

(ii) If X is reflexive, then X has the Property (P).

The monograph [5] is an excellent introduction to the theory of Banach spaces not containing ℓ^1. Here we show that a weaker version of the Property (P) is equivalent to X not containing ℓ^1, while among a certain class of Banach spaces, Property (P) implies reflexivity. Specifically we prove :

Theorem 4 *Let X be a Banach space.*

(a) X does not contain a copy of ℓ^1 if and only if every norm bounded, weakly sequentially complete *convex set in X has Property (P).*

(b) If X is weakly sequentially complete, then X is reflexive if and only if X has Property (P).

Proof : (a) Since ℓ^1 is weakly sequentially complete (see [6]), so is any closed convex subset of it, in particular, so is the set K in our example of Section 1. Note that if X contains ℓ^1, the set K above can be identified as a norm bounded, weakly sequentially complete convex subset of X which, by Section 1, lacks Property (P).

Conversely, if X does not contain a copy of ℓ^1 and $K \subseteq X$ is weakly sequentially complete then K is weakly compact. Indeed, if $\{x_n\}$ is a sequence in K, then by Rosenthal's ℓ^1 Theorem (see [5] or [4, Chapter XI]), it has a weak Cauchy subsequence which, by weak sequential completeness, is weakly convergent. And hence, by Eberlein-Smulian Theorem, K is weakly compact and, by Corollary 3, K has the Property (P).

The proof of (b) is essentially the same. ∎

REMARK : Can one relax the assumption of weak sequential completeness in any of the two statements above ?

ACKNOWLEDGEMENT : The author is indebted to his supervisor, Prof. A. K. Roy, for his continuous encouragement and help in this work. He also wishes to thank Dr. D. P. Sinha of Delhi University for pointing out the characterisation of ℓ^1 given in Section 3.

References

[1] E. M. Alfsen, *Compact Convex Sets and Boundary Integrals*, Ergebnisse der Mathematik und ihrer Grenzgebiete, Band **57**, Springer-Verlag, (1971).

[2] R. D. Bourgin, *Geometric Aspects of Convex Sets with the Radon-Nikodým Property*, Lecture Notes in Math., No. **993**, Springer-Verlag (1983).

[3] G. Choquet, *Lectures on Analysis*, Vol. II, W. A. Benjamin, Inc. New York (1969).

[4] J. Diestel, *Sequences and Series in Banach Spaces*, Graduate Texts in Math., No. **92**, Springer-Verlag (1983).

[5] D. van Dulst, *Characterisations of Banach Spaces containing* ℓ^1, CWI Tract, Amsterdam (1989).

[6] N. Dunford and J. T. Schwartz, *Linear Operators*, Vol. I, Interscience, New York (1958).

[7] B.-L. Lin, P.-K. Lin and S.L. Troyanski, *A Characterization of Denting Points of a Closed Bounded Convex Set*, Longhorn Notes, The University of Texas at Austin, Functional Analysis Seminar (1985–86), 99–101.

[8] B.-L. Lin, P.-K. Lin and S.L. Troyanski, *Characterizations of Denting Points*, Proc. Amer. Math Soc. **102** (1988), 526–528.

[9] R. R. Phelps, *Lectures on Choquet's Theorem*, Van Nostrand Math. Studies, No. **7**, D. Van Nostrand Company, Inc., (1966).

Stat–Math Division
Indian Statistical Institute
203, B. T. Road
Calcutta 700 035
INDIA

Boundedly Complete Bases in Various Locally Convex Spaces

P.K.JAIN,[1] A.M.JARRAH AND D.P.SINHA[2]

Dedicated to the memory of U.N. Singh

1. INTRODUCTION

Boundedly complete bases were essentially introduced by James [9] for a Banach space. The concept has also been studied under the name `γ - complete bases´, given by Kalton [10], for general locally convex spaces. For a comprehensive account of the work carried out in this direction, one may refer to [2,13,18].

In this paper, we study whether a boundedly complete basis in a Hausdorff locally covex space is retained under a change in the topology of the space. We show that generally this is not true. If $\{x_n\}$ is boundedly complete Schauder basis (b.c.-S.b.) for a Hausdorff locally convex space (l.c space) (X,T), then in a courser compatible topology on X, $\{x_n\}$ remains a b.c.-S.b., which is not generally true for a topological basis (t.b.). Further, if $\delta^X \subset \mu$ then the boundedly completeness of a S.b.$\{x_n\}$ is retained in a compatible finer topology. Also, the b.c.-S.b. are identified amoung the bounded multiplier -S.b. For, any S.b.$\{x_n\}$ in an S-space (X,T), it is shown that under a certain condition the associated sequence of coefficient functionals $\{f_n\}$ to $\{x_n\}$, is a b.c.-S.b. in the Mackey topology of dual space X^*. Furthermore, some conditions are obtained for a b.c.-t.b. in a l.c. topology. It is observed that under these conditions some of the results for a t.b. may not be true for a S.b. Numerous examples and counter examples have been exhibited to bring out whether the condition imposed to prove certain results in the paper are necessary or otherwise.

---

(1) The author is partially supported by the UGC(India).

(2) The author is supported by the CSIR (India).

2. BASIC CONCEPTS

Throughout X is a linear space and T a Hausdorff locally convex (l.c.) topology on X. By $X^{\neq}$ and X^{*} we denote, respectively the algebraic and topological dual of Y The symbols β, τ and σ shall denote, respectively, the strong, Mackey and weak topology of the indicated dual pair. By X^{+} and X^{b} we mean the sequential dual and the set of all bounded linear functionals on X, respectively, defined by

$$X^{+} = \{ f \in X^{\neq} : f(x_n) \to 0 \text{ for all T-null } \{x_n\} \text{ in } X\},$$

$$X^{b} = \{ f \in X^{\neq} : \sup_{x\in B} |f(x)| < \infty \text{ for all T-bounded } B \subset X \}.$$

Then, $X^{*}\subset X^{+}\subset X^{b}\subset X^{\neq}$. We write T^{+}(and T^{b})for the finest l.c. topology on X such that T and T^{+} have the same convergent sequences (respectively, T and T^{b} have the same bounded sets). For the actual definitions of T^{b} and T^{+}, one may refer to [14,19]. A l.c. space is said to be a Mazur space if $X^{*}= X^{+}$. A l.c. space is said to be an S-space if X^{*} is $\sigma(X^{*},X)$-sequentially complete. Throughout, any l.c. topology that we consider, will be Hausdorff and T_1 and T_2 are any two such topologies on X.

A sequence $\{x_n\}$in a l.c. space (X,T) is said to be a topological basis (t.b.), if for every x in X, there is a unique sequence $\{\alpha_n\}$ of scalars such that $x = \text{T-}\lim_{n\to\infty} \sum_{i=1}^{n} \alpha_i x_i$. One may associate a unique sequence $\{f_n\}$ in $X^{\neq}$ such that $f_n(x) = \alpha_n$, for all x. If for a t.b. $\{f_n\}$ in $X^{\neq}$, then it is said to be a Schauder basis (S.b.) In this case the sequence $\{f_n\}$ is said to be the associated sequence of coefficient functions (a.s.c.f.) to the S.b.$\{x_n\}$.

Let $\{x_n\}$ be a S.b. of (X,T) with the a.s.c.f.$\{fn\}$. Then, (i) $\{x_n\}$ is unconditional if the series $\Sigma_{i=1}^{\infty} f_i(x)x_i$ is unconditionally convergent for each x in X, (ii) conditional if it is not unconditional, (iii) bounded multiplier (b.m.-S.b.) if for each x in X and each sequence of scalars $\{\alpha_n\}$ such that $|\alpha_n| \le |f_n(x)|$ for all n,it follows that the series $\sum_{i=1}^{\infty} \alpha_i x_i$ converges in X and (iv) shrinking if $\{f_n\}$ is a S.b. for X^* in $\beta(X^*,X)$. A sequence $\{x_n\}$ in a l.c. space (X,T) is said to be minimal if th re is a sequence $\{f_n\}$ in X^* such that $f_i(x_j) = \delta_{ij}$, for all i and j. Further, $\{x_n\}$ is said to be w-linearly independent (w-l.i.) if whenever $T\text{-}\sum_{i=1}^{\infty} \alpha_i x_i = 0$, we have $\alpha_i=0$, for all i. Clearly, every S.b. is minimal and every minimal sequence is w-l.i.

Let w and ϕ denote, respectively, the spaces of all scalar valued sequences and all finitely non-zero scalar sequences. A non-trival vector subspace λ of w is said to be a sequence space. The Kothe dual λ^x of a sequence space λ is defined by

$$\lambda^x = \{\{\alpha_i\}\in w : |\Sigma_{i=1}^{\infty} \alpha_i \beta_i| < \infty, \text{ for all } \{\beta_i\}\in\lambda\}.$$

If $\phi\subset\lambda$, then $\langle\lambda,\lambda^x\rangle$ forms a dual pair and the condition $\phi \subset \lambda$ is necessary for this duality ([15], p.407).

A sequence space λ is said to be normal, if $\{\beta_i\}\in\lambda$ whenever $|\beta_i| \le |\alpha_i|$ for all i and for some $\{\alpha_i\}\in\lambda$, and perfect if $\lambda=\lambda^{xx}$. Let $\{\alpha_i\}\in\lambda$ and $M\subset\lambda$. Then $\{\alpha_i\}$ is said to dominate M provided $|\beta_i| \le |\alpha_i|$ for each $\{\beta_i\}$ in M and all i. Further, if T is any topology on λ compatible with the dual pair $\langle\lambda,\lambda^x\rangle$, then λ is said to be simple if every T-bounded set in λ is dominated by a point in λ. For a S.b.$\{x_n\}$ in a l.c. space (X,T) with the a.s.c.f. $\{fn\}$ in X^*, we define the associated sequence spaces δ and μ as follows:

$$\delta = \{\{f_n(x)\} : x \in X\},$$

$$\mu = \{\{f(x_n)\} : f \in X^*\}.$$

Then $<\delta,\mu>$ forms a dual pair ([13], page 52} under the bilinear form

$$<\{f_n(x)\},\{f(x_n)\}> = \sum_{n=1}^{\infty} f_n(x)f(x_n) = f(x).$$

The map $F : \delta \to X$ and $G : X^* \to \mu$ defined by $F(\{f_n(x)\}) = x$ and $G(f) = \{f(x_n)\}$ respectively, are bijective and continuous.

Throughout the sequel T_1 and T_2 shall denote two locally convex topologies on X satisfying $T_1 \le T_2$.

3. BOUNDEDLY COMPLETE SCHAUDER BASES

Let (X,T) be a l.c. space. A t.b.(S.b.) $\{x_n\}$ in X is said to be a T-boundedly complete topological basis (respectively, T-boundedly Schauder basis), denoted by T-b.c.-t.b. (respectively, T-b.c.-S.b.), if for every sequence $\{\alpha_n\}$ of scalars, the series $\sum_{n=1}^{\alpha} \alpha_n x_n$ is T-convergent, whenever the sequence $\{\sum_{n=1}^{k} \alpha_n x_n\}$ is T-bounded. We begin by showing that if T_1 and T_2 are two l.c. topologies on X with $T_1 \le T_2$, then a T-b.c.-S.b. may not be T_2-b.c.-S.b. and conversely.

EXAMPLE 3.1. Let J be the James space with the usual norm topology ([18], p.273) and $X = J^*$. Let $\{x_n\}$ be a b.c.-S.b. of J. Since, J is a quasi-reflexive Banach space of order 1, if π is the natural imbedding of J into J^{**}, then $\text{codim}_{J^{**}}\pi(J)=1$. Let $\phi \in X^*(=J^{**})$ such that $\phi \notin \pi(J)$ and $\{\phi_n\}$ in X^* be defined by $\phi_1=\phi$ and $\phi_n = \pi(x_{n-1})$, n=2,3,.... It is easy to verify that $\{\phi_n\}$ is a b.c.-S.b. of X^* in $T_2=\beta(X^*,X)$, in fact, the norm topology of X^*. However, since $\pi(J)$ is $T_1=\sigma(X^*,X)$-dense in X^*, $\{\phi_n\}$ is not a T1-S.b. of X^*.

EXAMPLE 3.2. Let $X = l^\infty$, $Y = l^1$ and $\{e_n\}$ the usual sequence of unit vectors. Then $\{e_n\}$ is a $T_1 = \sigma(X,Y)$-b.c.-S.b. ([5], Theorem I.2.2), while it is not even a $T_2 = \beta(X,Y)$-S.b. since the span of $\{e_n\}$ is not T_2-dense in X. However, $\{e_n\}$ is a T_2-Schauder basic sequence in X which is still not T_2-b.c. Indeed,

$\{\sum_{i=1}^{n} e_i\}$ is not T_2-Cauchy though it is T_2-bounded.

Note that in the Examples 3.1 and 3.2, the topologies T_1 and T_2 are not compatible. However, for compatible topologies, we have

THEOREM 3.3. Let T_1 and T_2 be two compatible l.c. topologies on X with $T_1 \leq T_2$. Then, every T_2-b.c.-S.b. is a T_1-b.c.-S.b.

We omit the proof of this result since we shall give a more general result in Corollary 4.4 (see Remark 4.5).The converse of Theorem 3.3 is not true in general, as we see in the next example.

EXAMPLE 3.4. Consider the sequence $\{x_n\} \subset c_o$ and $\{f_n\} \subset l^1$ defined by

$$x_n = \sum_{i=1}^{n} e_i \quad (n = 1,2,\ldots),$$

$$f_n = e_n - e_{n+1} \quad (n = 1,2,\ldots).$$

Then, $\{x_n\}$ is a non-shrinking S.b. of c_o in the norm topology with the a.s.c.f. $\{f_n\}$. Hence, $\{f_n\}$ is a $\sigma(l^1,c_o)$-b.c.-S.b. of l^1([20], Proposition 1.5). If possible, let $\{f_n\}$ be a $\tau(l^1,c_o)$-b.c.-S.b. Then, for any sequence $\{\alpha_n\}$ of scalars, if $\{\sum_{i=1}^{n} \alpha_i x_i\}$ is $\beta(l^1,c_o)$-bounded, it is $\tau(l^1,c_o)$-bounded and hence $\tau(l^1,c_o)$-convergent. Therefore, $\{\sum_{i=1}^{n} \alpha_i x_i\}$ is $\beta(l^1,c_o)$-convergent ([6], Theorem, 12), so that $\{f_n\}$ is a $\beta(l^1,c_o)$-b.c.-S.b. for the $\beta(l^1,c_o)$-closed linear span of $\{f_n\}$ in l^1, which is a contradiction ([18], Corollary 6.1(b),p.286).

Towards the converse of Theorem 3.3, we have

THEOREM 3.5. Let T_1 and T_2 be two compatible l.c. topologies on X with $T_1 \leq T_2$. If $\delta^x \subset \mu$, then a T_1-b.c.-S.b. is a T_2-b.c.-S.b.

PROOF. Let $\{x_n\}$ be a T_1-b.c.-S.b. Then, it is $\sigma(X,X^*)$-b.c.-S.b. by Theorem 3.3. Therefore, $\{e_n\}$ is a $\sigma(\delta,\mu)$-b.c.-S.b. of δ ([13], Theorem 3.3.5), whence δ is perfect ([13], Proposition 2.3.5). Therefore, under the hypothesis, we have $\delta^x = \mu$([12], Proposition 2.2.7 and [5], Theorem I.3.2). Hence, $(\delta,\tau(\delta,\delta^x))$ is barraled ([15], Section 30.7(1)), whence $\{x_n\}$ is a $\tau(X,X^*)$-S.b. ([1], Theorem 11). It is easy to verify that $\{x_n\}$ is a $\tau(X,X^*)$-b.c.-S.b. and hence by Theorem 3.3, a T_2-b.c.-S.b.

ALTERNATIVE PROOF If $\{x_n\}$ is a T_1-b.c.-S.b., then as in the begining of the first proof δ is perfect and so $\mu = \delta^x$. Therefore, $\{x_n\}$ is a T_1-b.m.-S.b. ([3], Theorem 6.1), whence is a T_2-b.m.-S.b. ([4], Theorem 3). This gives that $\{x_n\}$ is T_2-b.c.-S.b. ([3], Theorem 4.3).

REMARK 3.6. The condition $\delta^x \subset \mu$ in Theorem 3.5 is not necessary. Note that in case the b.c.-S.b. is conditional and $\delta^x \subset \mu$, then $\delta^x = \mu$. This contradicts Theorem III.1.6 in [5], and thus we must have $\delta^x \subset \mu$. However, we show that there is a conditional b.c.-S.b. satisfying the assertion of the theorem (see Example 3.7). Further, for an unconditional b.c.-S.b. also, we may have $\delta^x \subset \mu$, while the assertion of the theorem still holds. Indeed, for $X = l^1$, $T_1 = \sigma(l^1,c_o)$ and $T_2 = \tau(l^1,c_o)$, the unit vector basis $\{e_n\}$ of X is both T_1- and T_2-b.c.-S.b., while $\delta = l^1$, $\mu \doteq c_o$ so that $\delta^x \subset \mu$.

EXAMPLE 3.7. Consider the sequence $\{x_n\}$ and $\{f_n\} \subset l^2$ given by

$$x_{2n-1} = e_{2n-1} + \sum_{i=1}^{\infty} a_{i-n+1} e_{2i}, \quad x_{2n} = e_{2n},$$

$$f_{2n-1} = e_{2n-1}, f_{2n} = \sum_{i=1} (-a_{n-i+1}) e_{2i-1} + e_{2n},$$

for each n, where $\{a_n\}$ is such that $a_n \geq 0$ $(n \geq 1)$, $\sum_{j=1}^{\infty} j a_j^2 < \infty$ and $\sum_{j=1}^{\infty} a_j = \infty$. Then, $\{x_n\}$ is a $\beta(l^2,l^2)$-S.b. of l^2 with the a.s.c.f. $\{f_n\}$ such that $\{f_n\}$ is conditional ([16], Theorem). Hence, $\{f_n\}$ is a conditional $\sigma(l^2,l^2)$-b.c.-S.b. ([20], Proposition 1.5). Since the space l^2 with the norm topology is a reflexive Banach space, every $\sigma(l^2,l^2)$-convergent sequence is $\beta(l^2,l^2)$-convergent. One may verify that $\{f_n\}$ is boundedly complete as a $\beta(l^2,l^2)$-S.b.

It is observed in Theorem 6.1 in [3], that a b.c.-S.b. such that $\mu = \delta^x$ is a b.m.-S.b., of course, the S.b. in this case will be unconditional. Towards the converse of this resullt we have

THEOREM 3.8. Let $\{x_n\}$ be a b.m.-S.b. for a l.c. space (X,T) such that δ is simple. Then, $\{x_n\}$ is a $\sigma(X,X^*)$-b.c.-S.b.

PROOF. Since $\{x_n\}$ is a T-b.m.-S.b., it is $\sigma(X,X^*)$-b.m.-S.b. ([4], Theorem 3), whence $\{e_n\}$ is a $\sigma(\delta,\mu)$-b.m.-S.b. of δ ([13], Theorem 3.5.5). Hence, δ is normal ([13], Proposition 2.3.5). Now, δ being simple, is perfect ([7], Corollary 4). therefore, $\{e_n\}$ is a $\sigma(\delta,\mu)$-b.c.-S.b. for δ ([13], Proposition 2.3.5), so that $\{x_n\}$ is $\sigma(X,X^*)$-b..c.-S.b. ([13], Theorem 3.3.5).

The assumption that δ is simple in Theorem 3.8 can not be dropped as shown below.

EXAMPLE 3.9. Let $X = c_o$, equipped with the usual norm topology. Then, for the unit vector basis $\{e_n\}$ of X we have $\delta = c_o$, which is normal but not simple and perfect. Hence, $\{e_n\}$ is a b.m.-S.b. which is not a b.c.-S.b. ([5], Theorem I.2.5).

COROLLARY 3.10. Let $\{x_n\}$ be a b.m.-S.b. of a l.c. space (X,T) such that $\delta^x \subset \mu$ and δ is simple. Then $\{x_n\}$ is a T-b.c.-S.b.

Next, we show that the a.s.c.f. to a S.b. in an S-space, under certain conditions, is a b.c.-S.b. in the Mackey topology.

THEOREM 3.11. Let $\{x_n\}$ be a S.b. of an S-space (X,T) with the a.s.c.f. $\{f_n\}$ in X^* such that $\mu^x \subset \delta$. Then, $\{f_n\}$ is a $\tau(X^*,X)$-b.c.-S.b. of X^*.

PROOF. Since $\{f_n\}$ is a $\sigma(X^*,X)$-S.b. of X^* and $(X^*,\sigma(X^*,X))$ is sequentially complete, μ is perfect ([12], Theorem 2.5.1). Now since $\mu^x \subset \delta$, $\delta^x = \mu$. Therefore, $\{e_n\}$ is a $\sigma(\mu,\delta)$-b.c.-S.b. of μ([5], Theorem I.2.2). Hence, $\{f_n\}$ is a $\sigma(X^*,X)$-b.c.-S.b. of ([13], Theorem 3.3.5), whence by Theorem 3.5, $\{f_n\}$ is a $\tau(X^*,X)$-b.c.-S.b.

In Theorem 3.10 the condition $\mu^x \subset \delta$ is not necessary. Indeed, the a.s.c.f.$\{e_n\} \subset l^1$ of the unit vector basis of c_o with the norm topology, is a $\sigma(l^1,c_o)$-b.c.-S.b. Note that the space c_o is an S-space, $\delta = c_o$, $\mu = l^1$ and hence $\mu^x \subset \delta$. However. the condition for X to be an S-space, in Theorem 3.10 can not be dropped, consider the non-S-space $(l^1,\sigma(l^1,c_o))$ with the unit vector basis $\{e_n\}$. Then, we have $\delta = l^1$ and $\mu = c_o$ so that $\mu^x = \delta$. But the a.s.c.f$\{e_n\} \subset c_o$ is not $\sigma(c_o,l^1)$-b.c.-S.b.

REMARK 3.12. On the lines of the proof of Theorem 3.10, one may verify that for a weak sequentially complete l.c. space (X,T), every S.b.$\{x_n\}$ is T-b.c.-S.b. if $\delta^x \subset \mu$. Also, for any weak sequentially complete S-space, in particular, for any Mazur space, with a S.b.$\{x_n\}$ with the a.s.c.f.$\{f_n\}$, one may observe that $\{x_n\}$ and $\{f_n\}$ are T-b.c.-S.b. in X and X^*, respectively, if either $\delta^x \subset \mu$ or $\mathring{\mu}^x \subset \delta$.

4. BOUNDEDLY COMPLETE TOPOLOGICAL BASES

If T_1 and T_2 are compatible l.c. topologies on X with $T_1 \leq T_2$, we find in Theorem 3.3 that a T_2-b.c.-S.b. is a T_1-b.c.-S.b. But, this may not be true for t.b. (see Example 4.3). However, under certain conditions, this result, in fact, a rather more general result for a t.b., is also true.

THEOREM 4.1. Let T_1 and T_2 be any two l.c. topologies on X with $T_1 \leq T_2$. Let $\{x_n\}$ be a T_2--b.c.-t.b. of X such that it is T_1-w-l.i. and satifies the condition

(A) For every sequece $\{\alpha_n\}$ of scalars the sequence $\{\Sigma_{i=1}^{n} \alpha_i x_i\}$ is T_2-bounded whenever it is T_1-bounded.
Then, $\{x_n\}$ is T_1-b.c.-t.b.

PROOF. Since $\{x_n\}$ is a T_2-t.b. and T_1-w-l.i. it is T_1-t.b. Now, the T_1-bounded completeness of $\{x_n\}$ follows in view of condition (A).

COROLLARY 4.2. If $T_1 \leq T_2 \leq T_1^b$, then a T_2-b.c.-t.b. x_n is T_1-b.c.-t.b. if and only if $\{x_n\}$ is T_1-w-l.i.

The condition of w-linear independence in Theorem 4.1 can not be dropped.

EXAMPLE 4.3. Let X, T_1, T_2 and $\{x_n\}$ be as in Example 3.1. Since $\{x_n\}$ is a b.c.-S.b. in J, it is 1-shrinking ([17], Corollary 1). Thus, if $\{f_n\} \subset X = J^*$ is the a.s.c.f. to $\{x_n\}$, then $\text{codim}_X [f_n] = 1$. Hence, if $f_o \in X = J^*$ such that $f_o \notin [f_n]$, then $\{f_o, f_n\}$ is a basis of X. Let $\phi \in X^*$ such that $\phi(f_n) = 0$ for all n and $\phi(f_o) \neq 0$. Let $x \in J$ such that $f_o(x) = -\phi\,(f_o)$. write $\phi_o = \phi + \pi(x)$. Then, $\phi_o \notin \pi(J)$, since $\phi \notin \pi(J)$ and $\phi_o(f_o) = 0$. Let $\{\phi_n\} \subset X^*$ be defined by $\phi_1 = \phi_o$ and $\phi_n = \pi(x_{n-1})$, n=2,3,.... Then, $\{\phi_n\}$ is a T_2-b.c.-S.b. of X^*. Next, we show that $\{\phi_n\}$ is not $T_1 = \sigma(X^*,X)$-w-l.i. Let P be bounded linear projection of X onto $[f_n]$. Then, $P^*(\pi(x)) = \pi(x)$, for all $x \in J$. Now, for each $f \in X^*$, there exists a unique scalar β such that

$$f = \sum_{i=1}^{\infty} (Pf)(x_i) f_i + \beta f_o .$$

Then

$$\phi_1 = \sigma(X^*,X)- \sum_{i=1}^{\infty} \phi_1(f_i)\pi(x_i).$$

It follows that $\{\phi_n\}$ is not T_1-w-l.i. sequence.

Further, to justify our assertion in the brginbing of this section, let $T_3 = \tau(X^*,X)$ so that $T_1 \leq T_3$. Since $J^{**} = X^*$ is a norm-seperable Banach space, by the Corollary to Proposition 1 in [11], every $\tau(X^*,X)$-convergent sequence is $\beta(X^*,X)$-convergent. Hence, $\{\phi_n\}$ is $\tau(X^*,X)$-w-l.i. ([8], Proposition 3.4), whence by Corollary 4.2, $\{\phi_n\}$ is T_3-b.c.-t.b.

For a S.b., Theorem 4.1 takes the following form :

COROLLARY 4.4. Let T_1 and T_2 be two l.c. topologies on X with $T_1 \leq T_2$. If $\{x_n\}$ is a T_2-b.c.-S.b. of X such that it is T_1-minimal and satisfies condition (A). Then, $\{x_n\}$ is a T_1-b.c.-S.b.

REMARK 4.5. Note that if T_1 and T_2 are compatible topologies then, every T_2-minimal sequence is T_1-minimal and satisfies condition (A). Hence, Theorem 3.3 follows from Corollary 4.4.

THEOREM 4.6. Let T_1 and T_2 be two l.c. topologies on X with $T_1 \leq T_2$. If $T_2 \leq T_1^+$, then every T_1-b.c.-t.b. (T_1-b.c.-S.b.) is T_2-b.c.-t.b. (respectively, T_2-b.c.-S.b.).

PROOF. If $\{x_n\}$ is a T_1-b.c.-t.b. (T_1-b.c.-S.b.), then the fact that $\{x_n\}$ is T_2-t.b. (respectively, T_2-S.b.) and that it is T_2-b.c. follows by repeatedly using $T_2 \leq T_1^+$.

The above proof can be easily improved to give the following refinement of theorem 4.6.

THEOREM 4.7. Let $\{x_n\}$ be a b.c.-t.b. (b.c.-S.b.) in a l.c. space (X,T_1). Let T_2 be the finest l.c. topology on X such that $\{x_n\}$ is a T_2-t.b. (respectively, T_2-S.b.). Then, $\{x_n\}$ is a T_2-b.c.-t.b. (respectively, T_2-b.c.-S.b.).

REMARK 4.8. The condition $T_2 \leq T_1^+$ in Theorem 4.6 can not be totally dropped. Indeed, in Example 3.2 note that $X^\# = l^1$ ([15], section 30.1(4)), $T_1^+ \leq \tau(X,X^\#) < T_2$ ([19], Corollary 1.12) and $\{e_n\}$ is a T_1-b.c.-S.b., which is not a T_2-b.c.-t.b.

The converse of Theorem 4.6 may not be true for a S.b. Note that the T_3-b.c.-t.b. $\{\phi_n\}$ of X^* in Example 4.3 is not a T_3-S.b. since $\{\phi_n\}$ is not T_3-minimal. Further, $\{\phi_n\}$ is a T_2-b.c.-S.b. and $T_2 \le T_3^+$. However, for a t.b. the converse of Theorem 4.6 is true.

THEOREM 4.9. Let T_1^+ and T_2 be two l.c. topologies on X with $T_1 \le T_2$. If $T_2 \le T_1^+$, then a T_2-b.c.-t.b. is a T_1-b.c.-t.b.

PROOF. Note that $T_1 \le T_2 \le T_1^+ \le T_1^b$. If $\{x_n\}$ is a T_2-b.c.-t.b. of X, Then it is T_1-w-l.i. ([18], Proposition 3.4). Hence, by Corollary 4.2, $\{x_n\}$ is T_1-b.c.-t.b.

COROLLARY 4.10. Let λ be a sequence space such that λ^x is simple. Then every $\beta(\lambda,\lambda^x)$-b.c.-t.b. is $\sigma(\lambda,\lambda^x)$-b.c.-t.b.

One may observe that the condition $T_2 \le T_1^+$ in Theorem 4.9 can not dropped. Indeed, we only require to note that in Example 4.3, $T_1^+ < T_2$, for otherwise by Proposition 4.3 in [8], $\{\phi_n\}$ is T_1-w-l.i., which is not true.

REFERENCES

[1] M.G. Arsove and R.E.Edwards, Generalized basis in topological linear spaces, Studia Math. 19(1960), 95-113.

[2] M.M. Day, Normed linear spaces, Springer-Verlag, Berlin (1973).

[3] E. Dubinski and J.R. Retherford, Schauder bases and Kothe sequence spaces, Bull. Aca. Scs. (Ser. Sc. Math. Astr. Phy.) 16(9)(1966), 497-501.

[4] ______________________ Bases in compatible topologies, Studia Math. 28(1967), 221-226.

[5] ______________________ Schauder bases and Kothe sequence spaces, Trans. Amer. Math. Soc. 130(1968), 256-280.

[6] D.J.H.Garling, The β- and γ-duality, Proc. Camb. Phil. Soc. 63(1967), 963-981.

[7] R.T. Jacob, Matrix transformations involving simple sequence spaces, Pacific Jour. Math. 70(1)(1977), 179-187.

[8] P.K.Jain, A.M. Jarrah and D.P. Sihna, Minimal and w-linearly independent sequences in locally convex spaces (communicated).

[9]. R.C. James, Bases and reflexivity of Banach spaces, Ann. Math. 52(1950), 518-527.

[10] N.J. Kalton, Schauder bases and reflexivity, Studia Math. 38(1970), 255-266.

[11] ______________ Mackey duals and almost shrinking bases, Proc. Camb. Phil. Soc. 74(1973), 73-81.

[12] P.K. Kamthan and M.Gupta, Sequence Spaces and Series, Marcel Dekker, New York (1981).

[13] P.K. Kamthan and M.Gupta, Schauder bases-Behaviour and Stability, Longman, Harlow (1988).

[14]. J.L. Kelley, I. Namioka et. al., Linear topological spaces, Van Nostrand, Princeton (1963).

[15] G. Kothe, Topological Vector Spaces I, Springer-Verlag, Berlin (1969).

[16] B.L. Lin and I. Singer, On conditional bases of l^2, Ann. Soc. Math. Polon. Ser.I. Comm. Math. 15(1971), 135-139.

[17] I. Singer, Bases and quasi-reflexivity of Banach spaces, Math. Ann. 153(1964), 199-209.

[18] ____________ Bases in Banach Spaces I, Springer-Verlag, Berlin (1970).

[19] J.H. Webb, Sequential convergence in locally convex spaces, Proc. Camb. Phil. Soc. 64(1968), 341-364.

[20] L.J. Weill, Unconditional and shrinking bases in locally convex spaces, Pacific Jour. Math. 29(1969), 467-483.

P.K. Jain and D.P. Sinha
Department of Mathematics
University of Delhi
Delhi-110007
INDIA

A.M. Jarrah
Department of Mathematics
Jamia Millia Islamia
New Delhi-110025
INDIA

A REPRESENTATION OF THE MULTIPLIER MODULE $Hom_A(A,W)$

R. Vasudevan

DEDICATED TO THE MEMORY OF U.N. SINGH

The purpose of this paper is to establish sufficient conditions for the multiplier module $Hom_A(A,W)$ to be isometrically isomorphic to W, where W is a Banach left A-module over a Banach algebra A.

1. INTRODUCTION:

Let A be a Banach algebra and W a Banach left A-module over the Banach algebra A. Almost periodic Banach modules over commutative Banach algebras were first introduced by J.W. Kitchen, Jr in [3] and later R. Vasudevan in [5] considered the left and right adjoint of the restriction functor in the category of almost periodic Banach modules over Banach algebras. Wendel in [7] had shown that for a non-discrete locally compact group G, $Hom_{L_1(G)}(L_1(G), L_1(G)) = M(G)$ and hence in general $Hom_A(A,W)$ is not isometrically isomorphic to W. In this paper we establish sufficient conditions for $Hom_A(A,W) \cong W$.

DEFINITIONS AND PRELIMINARIES

DEFINITION 2.1 If A is a Banach algebra, then we call (W,*) a Banach left A-module (right A-module) if W is a Banach space and * is a map from A x V to V with the following properties.

(a) The map * from AxV to V is bilinear.

(b) $(a\cdot b) * w = a*(b*w)$, $a,b\in A$, $w\in W$.
$[(a\cdot b)*w = b*(a*w)]$

(c) The map * from AxV to V is continuous.

DEFINITION 2.2 Let A be a Banach algebra and W a Banach left A-module.
Then $Hom_A(A,W)$ is the Banach space of all continuous linear maps T from A to W satisfying $T(a*w) = a*T(w)$ for all $a\in A$ and $w\in W$ under the usual operator norm.

DEFINITION 2.3 Let A be a Banach algebra and W be a Banach left A-module. Then W is called an essential A-module if the vector space spanned by $\{a*w \mid a \in A, w \in W\}$ is dense in V.

REMARK 2.4 If A is a Banach algebra with bounded approximate identity, then viewed as a left or right Banach module over itself, A is an essential A-module.

DEFINITION 2.5 Let A be a Banach algebra and W a Banach left A-module. We say that an element $w \in W$ is almost periodic if and only if the mapping $T_w : a \to a*w$ from A to W is compact.

We say that the left A-module is almost periodic if each element in W is almost periodic.

REMARK 2.6 Although a Banach left A-module W need not itself be almost periodic, it does contain a largest closed submodule which is almost periodic, namely, the set, which we denote by $W a.p$, consisting of all the almost periodic elements of W. The fact that $W a.p$ is a closed submodule of W follows from the well-known properties of compact operators.

REMARK 2.7 $L_\infty(G)$∘ is a Banach left $L_1(G)$-module, where G is a locally compact group. Then the almost periodic elements of $L_\infty(G)$ are just the classical almost periodic functions on G.

REMARK 2.8 If W is a Banach left A-module then W^* is a right A-module under the usual module composition

DEFINITION 2.9 A Banach left A-module W is said to be order free if for each $w \in W$, $a*w = 0$ for all $a \in A$ implies $w = 0$.

REMARK 2.10 If A is a Banach algebra with bounded approximate identity, then any essential Banach left A-module is order free. If W is such a module, by Cohen-Hewitt factorization theorem[2],

$i_j * w \to w$ for every $w \in W$,

where $\{i_j\}_{j \in \wedge}$ is a two sided bounded approximate identity for A.

Hence $w = 0$.

REMARK 2.11 If G is a locally compact group, then $L_p(G)$, $1 \leq p \leq \infty$ is a Banach left $L_1(G)$-module. Further $L_p(G)$, $1 \leq p < \infty$ is essential and $L_\infty(G)$ is order free as a left $L_1(G)$-module.

3. We now establish our main result.

THEOREM 3.1 Let A a Banach algebra, which is almost periodic as a left A-module. Let W be a Banach left A-module. Then if the Banach A-module W^* is order free, then W is an almost periodic A-module.

PROOF. If W is not almost periodic, then $W \neq W_{a.p.}$. By Hahn-Banach Theorem, there exists $w_o^* \in W^*$, $w_o^* \neq 0$, such that $w_o^*(W_{\alpha.\rho}) = 0$. Since A is an almost periodic left A-module, we have for all a in A and $w \in W$, $a * w \in W_{a.p.}$

Thus

$$(a * W_o^*)(w) = w_o^*(a * w) = 0$$

for all $a \in A$ and $w \in W$, i.e. $a * w_o^* = 0$ for all $a \in A$.

Since W^* is order free, hence $w_o^* = 0$, a contradiction. Hence W is almost periodic as a Banach left A-module.

COROLLARY 3.2 Let A be a Banach algebra, almost periodic as a left A-module. Then if W is order free reflexive left A-module, W^* is almost periodic as a right A-module.

PROOF. $W = (W^*)^*$ is order free, hence by Th. 3.1, W^* is almost periodic left A-module

THEOREM 3.3. Let A be a Banach algebra with two-sided approximate identity of norm one. Let W be an essential left A-module. Then there is a natural isometric right module isomorphism

$$\mathrm{Hom}_A(A,(W^*)_{\alpha.\rho}) \cong (W^*)_{\alpha.\rho}.$$

PROOF. Let $z \in (W^*)\alpha.\rho$.

Define $T_z : A \to (W^*)\alpha.\rho$.

by $T_z(a) = a * z$

Since $z \in (W^*)\alpha.\rho.$, hence $a*z \in (W^*)\alpha.\rho$ and the mapping is from A to $(W^*)\alpha.\rho$.

Also $||T_z(a)|| = ||a * z|| \leq ||a||\ ||z||$

Hence $||T_z|| \leq ||z||$

Let $\epsilon > 0$ be given, choose $w \neq 0$, $||w||=1$ such that $||z(w)|| \geq ||z|| - \epsilon$. Since W is essential, by Cohen-Hewitt factorisation theorem [2] there exists $a \in A$, $w' \in W$ such that $a * w' = w$

We have $||(a*z)(w')|| = ||z(a*w'||$

$$= ||z(w)||$$

$$\geq ||z|| - \epsilon$$

Since ϵ is arbitrary, hence $||a*z|| \geq ||z||$

Thus the mapping $z \to T_z$ from $(W^*)\alpha.\rho$ to $\mathrm{Hom}_A(A,(W^*)\alpha.\rho)$ is an isometry.

We shall show that the map is onto. Let $T \in \mathrm{Hom}_A(A,(W^*)\alpha.\rho)$. Let $\{i_j\}_{j\in\Lambda}$ be a two-sided bounded approximate identity of norm one.

Then $\{T(i_j)\}_{j\in\Lambda}$ is a bounded net in $(W^*)\alpha.\rho$.

Hence there exists a bounded subnet $\{T(i_{j_k})\}_{k\in\Lambda'}$ of $\{T(i_j)\}_{j\in\Lambda}$ which converges in the weak*-topology of $(W^*)\alpha.\rho$.

Let $\lim_k T(i_{j_k})=z$, where limit is taken in the weak*-topology. Now the action of A on W^* is weak*-continuous. For if $w_j^* \to w_o^*$ in the weak*-top of W^*, then

$$(a * w_j^*)(w) = w_j^*(a * w) \to w_o^*(a*w)$$

$$= (a*w_o^*)(w).$$

i.e. $a * w_j^* \to a * w_o^*$ in the weak*-topology

Thus

$$T(a) = \lim_k T(a*i_{j_k})$$

$$= \lim_k \ a * T(i_{j_k})$$

$$= a * \lim T(i_{j_k})$$

$$= a * z \text{ for all } a \in A$$

Therefore $T = T_z$, where $z \in (W^*)a.p$

COROLLARY 3.4 If W is a reflexive order free Banach A-module, where A is a Banach algebra with two sided bounded approximate identity of norm one, then

$$Hom_A(A, Wa.p) \cong Wa.p$$

PROOF. $W = (W^*)^*$ is order free, hence W^* is an essential A-module. Hence by Theorem 3.3,
we have

$$Hom_A(A, [(W^*)^*]_{a.p}) \cong ((W^*)^*)a.p$$

that is

$$Hom_A(A, [W]_{a.p}) \cong Wa.p$$

COROLLARY 3.5 If W is an almost periodic reflexive order free left A-module, then

$$Hom_A(A,W) \cong W$$

COROLLARY 3.6 If A is a Banach algebra, which is almost periodic as a left A-module, then for any order free reflexive Banach left A-module W,

$$Hom_A(A,W) \cong W$$

PROOF $W = (W^*)^*$ is order free and hence W^* is essential A-module. Hence W^* is an order free module. Hence W is an almost periodic A-module. The result follows from Corollary 3.5

REFERENCES

1. Gulik, S.L., Liu, T.S. and Van Rooij, A.C.M, Group algebra modules I, Canad J. Math. 19, (1967), 130-150.

2. Hewitt, E and Ross, K.A., Abstract Harmonic Analysis, Vol I, Springer-Verlag, Berlin, (1963).

3. Kitchen, J.W., Jr. Normed modules and almost periodicity , Monat. für Math. 70 (1966), 233-243.

4. Rieffel, M.A., Induced Banach representations of Banach algebras and locally compact groups, J.Functional Analysis 1 (1967), 443-491.

5. R. Vasudevan, Almost periodic Banach modules and a representation of the multiplier module , Ph.D. Thesis, 1972.

6. Rudin, W.,Real and Complex Analysis, Tata-McGraw Hill, 1978.

7. Wendel, J.G., Left centralizers and isomorphisms of group algebras, Pac. J. Math. 2(1952), 251-261.

Department of Mathematics,
University of Delhi,
Delhi-110007.

ON TWO-PARAMETER SEMIGROUP OF OPERATORS

S.C. ARORA AND SHARDA SHARMA*

DEDICATED TO THE MEMORY OF U.N. SINGH

INTRODUCTION : For a Banach space X, a collection {T(t)} for each real number $t \geq o$ of bounded linear operators on the space X is said to be a semigroup of operators on X[1] if it satisfies the following:

i) $T(0) = 1$

ii) $T(t_1+t_2) = T(t_1)T(t_2)$ for $t_1, t_2 \geq 0$

The purpose of the present paper is to formulate the notion of two-parameter semigroup of operators, to define uniformly continuous and strongly continuous two-parameter semigroups and to study some of its properties. The paper has been divided into two sections. In Section 1, our aim is to study uniformly continuous semigroup whereas in Section 2, strongly continuous semigroups are defined and discussed.

For the space X, $\mathcal{B}(X)$ throughout, denotes the set of all bounded linear operators on X.

---

* The author gratefully acknowledges the support of U.G.C. grant No. F-26-1(2228)/(SR-IV) for this research work.

§1. UNIFORMLY CONTINUOUS SEMIGROUP:

Motivated by the definition of the semigroup, we introduce the following:

DEFINITION 1.1. A collection $A \subset B(X)$ is said to be a two-parameter semigroup of operators if

$$A = \{ T(t,s) : (t,s) \in [o, \infty[x[o,\infty[\}$$

where $T(t,s)$ is a bounded linear operator on X and the collection A satisfies

(i) $T(0,0) = I$

(ii) $T(t_1+t_2,s_1+s_2) = T(t_1,s_1)T(t_2,s_2)$

DEFINITION 1.2: A two-parameter semigroup $\{T(t,s)\}$ is said to be uniformly continuous semigroup if

$$\lim_{(t,s) \to (0,0)} || T(t,s) - I || = 0$$

This means that given any $\epsilon > 0$, there exists a $\delta > 0$ such that for all t and s satisfying $0<t<\delta$, $0<s<\delta$,

$$||T(t,s)-I|| < \epsilon.$$

We begin our study with the following

LEMMA 1.3: A two-parameter semigroup $\{T(t,s)\}$ is uniformly continuous if and only if

$$\lim_{(t,s) \to (t_0,s_0)} || T(t,s) - T(t_0,s_0) || = 0$$

for all $(t_0,s_0) \in [0,\infty[X[0,\infty[$.

PROOF: If

$$\lim_{(t,s) \to (t_0,s_0)} || T(t,s) - T(t_0,s_0) || = 0$$

is given then on putting $t_0 = 0 = s_0$, we get that $\{T(t,s)\}$ becomes uniformly continuous.

•On the other hand if $\{T(t,s)\}$ is uniformly continuous then for any (t_0,s_0) there exists K such that $\|T(t,s)\| \leq K$, for every t and s satisfying $0 \leq t \leq t_0 + 1$ and $0 \leq s \leq s_0 + 1$. The result follows by taking the limit of the difference.

DEFINITION 1.4: Generator:

For semigroup $\{T(t,s)\}$ let

$$M = \{ x \in X: \lim_{(t,s) \to (0,0)} \frac{T(t,s)x - x}{t+s} \text{ exists in } X \}$$

Define a map A with domain M as

$$Ax = \lim_{(t,s) \to (0,0)} \frac{T(t,s)x - x}{t+s}$$

This linear operator A is defined to be generator of the two-parameter semigroup $\{T(t,s)\}$.

Our next result indicates that a bounded linear operator generates a uniformly continuous two-parameter semigroup.

THEOREM 1.5: Any bounded linear operator A on the space X generates a uniformly continuous two-parameter semigroup of bounded linear operators on the space.

PROOF: For the bounded operator A, define

$$T(t,s) = e^{(t+s)A} = \sum_{n=o}^{\infty} \frac{((t+s)A)^n}{n!}$$

Since A is bounded so T(t,s) is a well-defined bounded linear operator on the space X for each $t \geq 0$ and each $s \geq 0$.

Also

$$\text{(i)}\quad T(0,0) = I$$
$$\text{(ii)}\quad T(t_1+t_2, s_1+s_2) = T(t_1,s_1)T(t_2,s_2)$$

Hence {T(t,s) is a semigroup of operators on the space X.

Also

$$\| T(t,s)-I \| \leq (t+s) \|A\| e^{(t+s)\|A\|}$$

therefore {T(t,s)} is uniformly continuous. To prove that A is the generator of {T(t,s)} we see that

$$\left\| \frac{T(t,s) - I}{t+s} - A \right\| \leq \|A\| [e^{(t+s)\|A\|} - 1]$$

Thus $\dfrac{T(t,s) - I}{t+s}$ converges to A in norm as $(t,s) \to (0,0)$ and hence strongly to A. Therefore A is the generator of {T(t,s)}.

Our next result establishes the uniqueness of the uniformly continuous semigroup corresponding to a fixed bounded linear operator as a generator.

THEOREM 1.6: Let {T(t,s)} and {S(t,s)} be two uniformly continuous semigroups of bounded linear operators on X. Let A be a bounded linear operator on X satisfying

$$\lim_{(t,s) \to (0,0)} \frac{T(t,s) - I}{t+s} = A = \lim_{(t,s) \to (0,0)} \frac{S(t,s) - I}{t+s}$$

then T(t,s) = S(t,s) for all (t,s) in $[0, \infty[\times [0,\infty[$.

PROOF: Let (t_0,s_0) be any arbitrary fixed ordered pair taken in $[0,\infty[\times [0,\infty[$. Since the mappings $(t,s) \to ||T(t,s)||$ and $(t,s) \to ||S(t,s)||$ are continuous mappings and hence are bounded on the rectangle $[0, t_0; 0,s_0]$ and therefore there exists a real number $K \geq 1$ satisfying

$$||\, T(t,s)\, ||\ ||\, S(t,s)\, || \leq K$$

for every (t,s) in $[0,t_0; 0,s_0]$. Now let $\epsilon > 0$ be any real number. Then by hypothesis there exists a $\delta > 0$ such that

$$\frac{1}{t+s}\, ||\, T(t,s)-S(t,s)\, || = \left|\left|\frac{T(t,s)-I}{t+s} - \frac{S(t,s)-I}{t+s}\right|\right|$$

$$< \frac{\epsilon}{K(t_0+s_0)}$$

whenever $0 < t < \delta$, $0 < s < \delta$. Choose a positive integer m such that $\frac{t_0}{m} < \delta$, $\frac{s_0}{m} < \delta$. Then

$$||\, T(t_0,s_0)-S(t_0,s_0)\,|| = \left|\left| \sum_{k=0}^{m-1} T\left(\frac{m-k}{m}t_0, \frac{m-k}{m}s_0\right) S\left(\frac{kt_0}{m}, \frac{ks_0}{m}\right) \right.\right.$$

$$\left.\left. -T\left(\frac{m-k-1}{m}t_0, \frac{m-k-1}{m}s_0\right) S\left(\frac{(k+1)}{m}t_0, \frac{(k+1)}{m}s_0\right)\right|\right|$$

$$\leq \sum_{k=0}^{m-1} \left|\left| T\left(\frac{m-k-1}{m}t_0, \frac{m-k-1}{m}s_0\right)\right|\right|\ \left|\left|S\left(\frac{kt_0}{m}, \frac{ks_0}{m}\right)\right|\right|$$

$$\left|\left| T\left(\frac{t_0}{m}, \frac{s_0}{m}\right) - S\left(\frac{t_0}{m}, \frac{s_0}{m}\right)\right|\right|$$

$$< K.\ \frac{t_0+s_0}{m} \cdot \frac{\epsilon}{K(t_0+s_0)} \cdot m$$

$$= \epsilon$$

This is true for each $\epsilon > 0$. Hence $T(t_0,s_0) = S(t_0,s_0)$.

Therefore $T(t,s) = S(t,s)$ for each (t,s) in $[0,\infty[\times [0,\infty[$.

§ 2. STRONGLY CONTINUOUS SEMIGROUP:

DEFINITION 2.1. A two-parameter semigroup $\{T(t,s)\}$ on X is said to be strongly continuous if for all x in X.

$$\lim_{(t,s)\to(0,0)} \| T(t,s)x-x \| = 0.$$

THEOREM 2.2: Let $\{T(t,s)\}$ be a strongly continuous semigroup of operators on X. Then there exists $M \geq 1$ and $W \geq 0$ such that for every (t,s) in $[0,\infty[\times [0,\infty[$.

$$\|T(t,s)\| \leq M e^{W(t+s)}$$

PROOF. We claim that there exists a real number $\eta > 0$ such that $\{T(t,s)\}$ is bounded on the square $[0,\eta;\ 0,\eta]$. If it is not so then for each natural number n, we can find real numbers t_n and s_n such that

$$0 < t_n < 1/n$$

$$0 < s_n < 1/n$$

and $\|T(t_n, s_n)\| > n$. Thus $(t_n, s_n) \to (0,0)$ and hence for each x in X, $\langle T(t_n, s_n)\rangle_{n=1}^{\infty}$ is convergent and hence bounded and therefore by Uniform Boundedness Principle, $\langle \|T(t_n, s_n)\| \rangle_{n=1}^{\infty}$ is a bounded sequence contradicting $\|T(t_n, s_n)\| > n$ for each n. Hence there exists $\eta > 0$ and M such that

$$\|T(t,s)\| \leq M$$

if $0 \leq t \leq \eta$ and $0 \leq s \leq \eta$. Since $\|T(0,0)\| = \|I\| = 1$, therefore $M \geq 1$. Now for any (t,s) in $[0,\infty[\times [0,\infty[$ we can find integers m and n such that $t = m\eta + \delta_1$ and $s = n\eta + \delta_2$ where $0 \leq \delta_1 < \eta$ and $0 \leq \delta_2 < \eta$. Let $w = 1/\eta \log M$. Then

$$\begin{aligned}\|T(t,s)\| &= \|T(m\eta, n\eta)\, T(\delta_1, \delta_2)\| \\ &\leq M^{n+m+1}\end{aligned}$$

$$\le M.\, M^{\frac{t+s}{\eta}}$$

$$\le M\, e^{w(t+s)}$$

and this proves the theorem.

As a consequence, we obtain the following.

COROLLARY 2.3: If $\{T(t,s)\}$ is a strongly continuous semigroup of operators on the space X then for each x in X, the mapping $(t,s) \rightarrow T(t,s)x$ is a continuous function from $[0,\infty[\ X\ [0,\infty[$ into X.

The authors are thankful to Prof. B.S.Yadav for his help in the preparation of this work.

REFERENCES

1. A.Pazy: Semigroup of Linear Operators and Applications to Partial differential Equations. Springer-Verlag, N.Y., 1983

DEPARTMENT OF MATHEMATICS
UNIVERSITY OF DELHI
DELHI - 110 007
INDIA.

HIGHER FRÉCHET AND DISCRETE GÂTEAUX DIFFERENTTIABILITY OF n-CONVEX FUNCTIONS ON BANACH SPACES

SAHEB DAYAL

DEDICATED TO THE MEMORY OF U.N. SINGH

1. INTRODUCTION It is well known that a convex function defined on a Banach space has a one-sided Gâteaux differential, see Flett [7;p. 253] and Kolony [14, p. 743]. We introduce k-discrete differences to generalize the notion of divided differences for a vector valued function, and consequently, define an n-convex function. It is shown that an n-convex locally bounded function admits a strong (n-2)-taylor series expansion. The kth coefficient of this series is a k-discrete Gâteaux differential, $0 \leq k \leq n-2$. It is further shown that such a function posses a one-sided (n-1)-discrete differential about every point of its domain. If, in addition, the one-sided (n-1)-discrete differential is continuous at a point, then it is proved that each coefficient of (n-2) strong Taylor series has a strong Taylor series expansion. The main result Theorem 3.3 establishes that if an n-convex locally bounded function admits a strong n-Taylor series expansion at a point whose coefficients are bounded multilinear operators, then it has nth Fréchet differential at that point.

2. PRELIMINARIES

Higher Fréchet Differentials : The meaning of Fréchet differentiability and higher Fréchet differentiability in the inductive sense is elegantly presented by Diéudonne [6] (see also [4]). A function $f : A \to F$ is said to be Fréchet differentiable at $x \in A$ if there exists a continuous linear transformation $\lambda_x \in L(E,F)$ the space of bounded linear transformation from E to F with the property that for all $\in > 0$ there is a $\delta > 0$ such that $|t| < \delta$ implies $||f(x+tu)-f(x)- \lambda_x(tu)|| < \in.|t|$ uniformly with respect to $u \in E$, $||u|| \leq 1$. If f is Fréchet differentiableat every point $x \in A$ it determines a vector valued function $f' : A \to L(E,F)$ by the formula $f'x = \lambda_x$. Thus the Fréchet differential of f at x is denoted by $f'x$. The Fréchet differentiation process can be iterated. If $f : A \to F$ is Fréchet differentiable in A the vector valued function $f' : A \to L(E,F)$ may again be Frechet differentiable in the above sense. Thus we have again a vector valued function $f'':A \to L(E,L(E,F))$ and in general $f^{(n)}: A \to L(E,L(E,\ldots,L(E,F))$, which defines the higher Fréchet differential in inductive sense, that is, with reference to lower differential. Because of natural isometry we may regard $L(E, L(E,F))$ as equivalent to $L_2(E,F)$ the space of bounded bilinear operators from E^2 to F and in general $f^{(n)}:A\to L_n(E,F)$, $L_n(E,F)$ is space of bounded multilinear operators from E^n to F ([6] p. 78).

Discrete difference and n-convex function : Let F be any Banach space, a n-discrete difference ([5], p.78) $[\Delta_n h](t_0,t_1,\ldots,t_n)$ of a function $h : [a,b]\to F$ where $(t_0,t_1,\ldots,t_n)$ is a finite sequence of distinct numbers in $[a,b]$, is the coefficient of t^n in the unique polynomial $P(t)$ of degree $\leq n$ such that $P(t_i) = h(t_i)$ for $i = 0,1,\ldots,n$. Thus

(2.1) $$[\Delta_n h](t_0,t_1,\dots,t_n) = \sum_{k=0}^{n} [1/(\prod_{\substack{j\in[0,n]\\ j\neq i}} (t_i-t_j))].h(t_k)$$

In case when F is taken to be the field of real numbers a n-discrete difference is said to be <u>monotonically increasing</u> ([5],p.79) if

$$[\Delta_n h]\,(t_0,t_1,\dots,t_n) \le [\Delta_n h]\,(t_0^*,t_1^*,\dots,t_n^*)$$

for any finite sequences $(t_0,\dots t_n)$ and $(t_0^*,\dots,t_n^*)$ such that $t_i \le t_i^*$ for $i = 0,1,\dots,n$. Discrete difference notion ([4],p.266) is the extension to the vector valued function of the notion of divided difference of Numerical Analysis [10].

DEFINITION 2.1 Let A be an open subset of a Banach space E and $f : A \to R$. For every $y \in A$ and $v \in E$ we define a function $[h(y,v)]t = f(y+tv)$. Let $[\Delta_n h(y,v)]\,(t_0,t_1,\dots,t_n)$ be n-discrete difference of h(y,v) defined for any finite sequence $(t_0,t_1,\dots,t_n)$ with $t_i \neq t_j$, $i \neq j$ and such that $y + t_j v$ is in the domain of ffor all $j = 0,1,2,\dots,n$, $y \in A$ and $v \in E$. f is said to be <u>n-convex</u> if for all $y \in A$ and $v \in E$, the function $\Delta_n h(y,v)$ does not change sign as a function of $(t_0,t_1,\dots,t_n)$.

The n-discrete difference $\Delta_n h(y,v)$ associated with $f : A \to R$ is said to be <u>uniformly bounded in the δ-neighbourhood</u> N_δ of $y_o \in A$ if there exists a $\delta' > 0$ and a constant M such that whenever $|t_i| < \delta'$, $i = 0,1,\dots,n$, $y \in N_\delta$ and $||v|| \le 1$, then $y + t_i v \in A$ and $|[\Delta_n h(y,v)](t_0,t_1,\dots,t_n)| \le M$. The n-discrete difference is said to <u>be</u>

<u>strongly uniformly continuous in $\delta/2$-neighbourhood</u> $N_{\delta/2}$ of $y_o \in A$ if for a given $\epsilon > 0$ there is a $\delta'' > 0$ such that whenever $y \in N_{\delta/2}$, $|s_i| < \delta/4$, $||v|| \le 1$, $|t_i| < \delta/4$ and $\sup_{0 \le i \le n} |t_i - s_i| < \delta''$ then

$$|[\Delta_n h(y,v)](t_0,t_1,\dots,t_n) - [\Delta_n h(y,v)](s_0,s_1,\dots,s_n)| < \epsilon$$

Discrete Higher Differentials (Discrete Differentials)

Higher differentials can be defined directly with no reference to the lower differentials. Riemann (see Butzer and Kosakiewez [2]), Peano, (H,β)-Peano (see Weil [17], Ash [7]) and Taylor differentials (see Nashed [15]) are introduced to define higher differentiability directly. We introduce, one-sided k-discrete, k-discrete Gâteaux and k-discrete Fréchet differentials [5] by using the discrete difference.

DEFINITION 2.2 [5] Let E and F be two Banach spaces and A be an open subset of E. Let $[\Delta_k h(y,v)]$ $(t_0,t_1,\dots,t_k)$ be the k-discrete difference associated to $f:A \to F$ defined for any finite sequence $(t_0,t_1,\dots t_k)$ with $t_i \ne t_j$, $i \ne j$ and such that $y+t_j v$ is in domain of f for all $j = 0,1,\dots,k$ $y \in A$ and $v \in E$. Denoting $[\Delta_k h(y,v)]$ $(t_0,t_1,\dots,t_k)$ by $[\Delta_k h(y,v)]t$ with notation $t = (t_0,t_1,\dots,t_k)$, if suppose the limit

$$(2.2) \qquad (\hat{f}^{(k)}y)v = (k!) \lim_{t \to 0+} [\Delta_k h(y,v)]t$$

exists for all $v \in E$ in the sense that for a given $\epsilon > 0$ there is a $\delta > 0$ such that for all $t = (t_0,t_1,\dots,t_k)$ with $0 < t_i < \delta$, $t_i \ne t_j$ $i \ne j$, $i,j = 0,1,2,\dots,k$.

(2.3) $\quad |(k!)\,[\Delta_k h(y,v)]t - (\hat{f}^{(k)}y)v| < \epsilon.$

Then $(\hat{f}^k y)$is called the <u>one-sided</u> <u>k-discrete</u> <u>differential</u>

of f at y. In case the choice of δ is such that $\delta > 0$ and for all $t = (t_0, t_1, \ldots, t_k)$, $|t_i| < \delta$, $t_i \neq t_j$, $i \neq j$, $i,j = 0,1,2,\ldots,k$

(2.3) holds, then $(f^{(k)}y)$ is called <u>k-discrete</u> <u>differential</u> of f at y. f is said to have a <u>k-discrete</u> <u>Gâteaux</u> <u>differential</u> $(f^{(k)}y) : E^k \to$

R if $(f^{(k)}y)$ is a k-linear operator such that $(\hat{f}^{(k)}y)u = (f^{(k)}y)(u,u,\ldots,u)$, provided limit (2.2) exists. In addition, if the limit (2.2) is uniform for vectors $v \in E$ such that $||v|| \leq 1$, then $(f^{(k)}y)$ is called the <u>k-discrete</u> <u>Fréchet</u> <u>differential.</u>

A one-sided 1-discrete differential, 1-discrete Gâteaux differential and 1-discrete bounded Fréchet differential are same as one-sided Gâteaux, Gâteaux and Fréchet differentials, respectively.

3. Local Representation and Higher Differentiability of n-Convex Functions

DEFINITION 3.1 [4]. Let E and F be any Banach spaces and A be an open subset of E. We say $f : A \to F$ has a <u>weak</u> <u>n-Taylor</u> <u>series</u> <u>expansions</u> about $y \in A$ if there are symmetric k-linear functions $P_k(y)$, possibly unbounded, such that

$$(3.1) \qquad f(y+tu) = \sum_{k=0}^{n} \frac{1}{k!} P_k(y)\,(tu)^k + r(u,t)$$

where $\lim_{t\to 0} r(u,t)/t^n = 0$.

If the limit is uniform for vectors u, $||u|| \leq 1$, then (3.1) is a <u>strong n-Taylor series expansion about</u> y.

THEOREM 3.1 Let E be a Banach space and A be an open subset of E. Let $f : A \to R$ be a locally bounded n-convex function, $n \geq 3$. Then there exists, for every y, bounded k-linear functions $(f^{(k)}y)$, $k = 0,1,2,3,\ldots,n-2$ such that f has a strong (n-2) Taylor series expansion and $(f^{(k)}y)$ $k = 0,1,2,\ldots,n-2$ are precisely the k-discrete Gâteaux differentials of f at y.

To prove Theorem 3.1 we use the following results which are already proved by the author in [5].

LEMMA 3.1 ([5] Theorem 3.1, p. 78): Let A be an open subset of Banach space E and $f : A \to R$ be n-convex. Then the (n-1) discrete difference $\Delta_{n-1}h(y,v)$ associated to f is monotonic. If, in addition f is bounded in a neighbourhood N_δ of yo, then $\Delta_{n-1}h(y,v)$ is uniformly bounded in $N_{\delta/2}$ and $\Delta_k h(y,v)$ is strongly uniformly continuous in $N_{\delta/2}$, $0 \leq k \leq n-2$.

LEMMA 3.2 ([5] Theorem 3.5, p. 81) : If A is any open subset of a Banach space E, $f : A \to R$ is n-convex, $n \geq 3$ and f is bounded in a neighbourhood N_δ of y_o then given $\epsilon > 0$ there is a $\delta > 0$ such that

(3.2) $$|[\Delta_1 h(y',v)]\,(0,t) - [\Delta_1 h(y'',v)]\,(0,t)\,| < \epsilon$$

whenever $||y'-y_o|| < \delta/4$, $||y''-y_o|| < \delta/4$ $||y'-y''|| < \delta'$, t is a real number with $0 < t < \delta$ and $||v|| \leq 1$.

LEMMA 3.3 ([5] Theorem 3.2 p. 80). Let A be any open subset of a

Banach space E and $f : A \to R$ be n-convex $(n > 2)$ locally bounded. Then

(i) For $0 \leq k \leq n-2$, the discrete differential $(\hat{f}^{(k)}y)$ exists and is bounded for every $y \in A$.

(ii) The one-sided (n-1)-discrete differential $(\hat{f}^{(n-1)}y)$ exists and is bounded for every $y \in A$.

(iii) f is uniformly continuous in a neighbourhood of every point $y \in A$.

LEMMA 3.4 ([5] Theorem 3.4 p. 81). Let A be any open subset of a Banach space E. If $f : A \to R$ is n-convex, $n \geq 2$, and bounded in a neighbourhood N_δ of y_o then for every $\epsilon > o$, there is a $\delta' > 0$ such that

$$|(\hat{f}^{(k)}y')v - (\hat{f}^{(k)}y'')v| < \epsilon$$

whenever $k = 0,1,2,\dots,n-2$, $y',y'' \in N_{\delta/4}$ $||y'-y''|| < \delta'$ and $||v|| \leq 1$.

REMARK 3.1 For any Banach space E and F and an operator $\alpha : E^n \to F$, E^n is n-cartesian product of E, one can naturally define an associated operator $\hat{\alpha} : E \to F$ defined by the formula $\hat{\alpha}u = \alpha(u,u,\dots,u)$. Less naturally for every operator $\beta : E \to F$ one may define an operator

$$(3.3) \qquad \tilde{\beta}(u_1,u_2,\dots,u_n) = \frac{1}{2^n n!} \sum_{(t_1,t_2,\dots,t_n)\in T} \left(\prod_{k=1}^{n} t_k\right)\beta(t_1u_1+\dots+t_nu_n$$

where T is the set of all finite sequence $(t_1,t_2,\dots,t_n)$ with $t_k = +1$ or -1 for every $1 \leq k \leq n$.

It can easily be seen that if $\alpha : E^n \to F$ is a n-linear symmetric operator then we can define $\hat{\alpha} : E \to F$ and hence $\tilde{\hat{\alpha}} : E^n \to F$ using (3.3), it turns out that $\tilde{\hat{\alpha}} = \alpha$. Also Taylor [16] proved that

$$(3.4) \qquad ||\hat{\alpha}|| \leq ||\alpha|| \leq \frac{n^n}{n!}||\hat{\alpha}||$$

(See also [4] p. 267). The inequality (3.4) is best possible proved by Kopeć and Musielak [11].

REMARK 3.2 For $n = 2$ Theorem 3.1 is still valid by Lemma 3.3 (iii) since only uniform continuity is claimed in this case.

PROOF OF THEOREM 3.1 For any y, v and $\eta > 0$, consider the polynomial

$$[P_\eta(y,v)]t = f(y) + \sum_{k=1}^{n-2} \left(\prod_{j=0}^{k-1}(t - j\eta)\right) . [\Delta_k h(y,u)]\ (0,\eta,2\eta,\dots,k\eta)$$

which coincides with $[h(y,u)]$ at $0,\eta,\dots,(n-2)\eta$. For t different from $0,\eta,\dots,(n-2)\eta$.

$$f(y+tu) = [P_\eta(y,u)]t + \left(\prod_{j=0}^{n-2}(t - j\eta)\right) . [\Delta_{n-1} h(y,u)]\ (0,\eta,\dots,(n-2)\eta,t)$$

Then,

$$(3.5)\quad f(y+tu)-[P_\eta(y,u)]t = \left(\prod_{j=0}^{n-2}(t - j\eta)\right).\frac{1}{t}\{[\Delta_{n-2}h(y,u)]\ (t,\eta,\dots,(n-2)\eta) - [\Delta_{n-2}h(y,u)]\ (0,\eta,\dots,(n-2)\eta)\}$$

$$f(y+tu)-[P_\eta(y,u)]t = \left(\prod_{j=0}^{n-2} (t - j\eta)\right). \{[\Delta_{n-2}h(y,u)]\ (t,\eta,\dots,(n-2)\eta) - [\Delta_{n-2}h(y,u)]\ (0,\eta,\dots,(n-2)\eta)\}$$

For any $t \neq 0$ we may choose η so small that $(n-2)\eta < |t|$ and $\eta \to 0$. The left hand side of (3.5) approaches

$$(3.6)\qquad f(y+tu) - \sum_{k=0}^{n-2} (t^k/k!)\ (\hat{f}^{(k)}y)u$$

by using Lemma 3.3 (i).

By Lemma 3.1 $[\Delta_{n-2}h(y,u)]$ is strongly uniforly continuous in $N_{\delta/2}$ neighbourhood of a point $y_o \in A$, the right hand side of (3.5) can be bounded by $\epsilon . t^{n-2}$, for t small and y near y_o and so

$$\lim_{t \to 0} [f(y+tu) - \sum_{k=0}^{n-2} (t^k/k!).(\hat{f}^{(k)}y)u]/t^{n-2} = 0$$

uniformly for $||u|| \leq 1$ and $y \in N_{\delta/2}$.

This gives

$$(3.7) \qquad f(y+tu) = \sum_{k=2}^{n-2} (t^k/k!)\ (\hat{f}^{(k)}y)u + r$$

such that $\lim_{t \to 0} r/\ t^{n-2} = 0$ uniformly for $||u|| < 1$ and $y \in N_{\delta/2}$.

The result will now follow quite easily if we show that if $1 \leq k \leq n-2$, there exists a bounded k-linear function $(f^{(k)}y)$ such that $(f^{(k)}y)u^k = (\hat{f}^{(k)}y)u$.

First we show that $(\hat{f}^{(k)}y)\ (su) = s^k(\hat{f}^{(k)}y)u$, if $k \geq 1$

If s =0, $[h(y,su)]t = fy$ independent of t and so

$$(\hat{f}^{(k)}y)\ (su)=0=s^k(\hat{f}^{(k)}y)u.$$

If $s > 0$, then

$$[\Delta_k h(y,sv)]\ (t_0,t_1,\dots,t_k) = s^k\ [\Delta_k h(y,u)]\ (st_0,st_1,\dots,st_k)$$

and the result follow from the limit as $t \to 0+$ (of. (2.2)). For s = -1 we have

$$[\Delta_k h(y,-u)]\ (t_0,t_1,\dots,t_k) = (-1)^k\ [\Delta_k h(y+t_k\ u,u)]\ (0,t_k-t_{k-1},..,t_k-t_0)$$

and continuity of $\hat{f}^{(k)}$(cf. Lemma 3.4) together with uniform approximation of $(\hat{f}^{(k)}y)u$ by values $(k!).\ [\Delta_k h(y,u)]\ t$ gives the result.

If $k = n-1$, the proof of homogeneity with respect to the non-negative scalars is still valid. If $\hat{f}^{(n-1)}$ is continuous at y then the proof can be modified rather simply to imply homogeneity of $f^{(n-1)}$ at y.

To complete the proof of linearity, we proceed by induction on k, starting with k = 1, considering explicitly only two terms on the right hand side of (3.7) we have

(3.8) $\quad f(y+tu) = f(y) + (\hat{f}^{(1)}y)(tu) + r,$

where $\lim_{t\to 0+} r/t = 0$ uniformly for $||u|| \le 1$. Similarly we obtain expressions for f(y+tu) and f(y+tu+tv) and combining we obtain

$$(3.9) \quad (\hat{f}^{(1)}y)(u+v) - (\hat{f}^{(1)}y)u - (\hat{f}^{(1)}y)v = (1/t)\{f(y+t(u+v)) - f(y+tu) - (y+tv) + fy\} + r/t$$

$$= [\Delta_1 h(y+tu,v)]\ (0,t) - [\Delta_1 h(y,v)]\ (0,t) + r/t$$

where r is a combination of remainders from three expansions and satisfies the condition

$$\lim_{t\to 0+} r/t = 0, \text{ uniformly for } ||u|| \le 1 \text{ and } ||v|| \le 1.$$

Using Lemma 3.2 the expression on the right of (3.9) tends to zero uniformly for $||u|| \le 1$, $||v|| \le 1$ and y near y_o, so,

$(\hat{f}^{(1)}y)\ (u+v) = (\hat{f}^{(1)}y)u + (\hat{f}^{(1)}y)v$, provided $||u|| \le 1$ and $||v|| \le 1$ and since we have shown that $(\hat{f}^{(1)}y)$ is homogeneous, it is linear. Of course, we define $(f^{(1)}y)$ to be $(\hat{f}^{(1)}y)$.

Inductively suppose bounded multilinear functions $(f^{(1)}y),\ldots,(f^{(k-1)}y)$ have been determined to give the values of $(\hat{f}^{(1)}y),\ldots,(\hat{f}^{(k-1)}y)$. Then we consider the estimation of f(y+su+stv) in two ways, by expansion about y and by expansion about y+su, stopping with the kth term.

$$f(y+su+stv) = \sum_{j=0}^{k} (1/j!)\ (\hat{f}^{(j)}y)\ (su+stv) + r_1$$

$$= \sum_{j=0}^{k} (1/j!)\ (\hat{f}^{(j)}\ (y+su))(stv) + r_2.$$

Solving $(\hat{f}^{(k)}y)$ (su+stv) we have

$$(\hat{f}^{(k)}y)(su+stv) = (k!)\ [\sum_{j=0}^{k} (1/j!)\ (\hat{f}^{(j)}(y+su))\ (stv)$$

$$- [\sum_{j=0}^{k-1} (1/j!)\ (f^{(j)}y)\ (su+stv)^j) + r_2 - r_1].$$

Then

$$(\hat{f}^{(k)}y)(u + tv) = (k!/s^k)\ [\sum_{j=0}^{k} (1/j!)\ (\hat{f}^{(j)}(y+su))\ (stv)$$

$$- [\sum_{j=0}^{k-1} (1/j!)\ (f^{(j)}y)\ (su+stv)^j) + r_2 - r_1].$$

Since $(f^{(j)}y)$is multilinear for $j \leq k-1$, and $(f^{(j)}(y+su))$is homogeneous. for $j \leq k$, the terms on the right, except for $r_2 - r_1$ form a polynomial in t. Since $\lim_{s \to 0+} (r_2 - r_1)/s^k = 0$ and the limit

is uniform for $|t| \leq 1$ for u and v fixed, $(\hat{f}^{(k)}y)\ (u+tv)$ is uniform limit of polynomials of degree $\leq k$ in t, if $|t| \leq 1$ and hence $(\hat{f}^{(k)}y)$ $(u+tv)$ is itself a polynomial of degree $\leq k$ in t, if $|t| \leq 1$.

Let $u_1, u_2, \ldots, u_r$ be a basis of a finite dimensional subspace of E. Putting $t_1u_2 + \ldots + t_ru_r = u+tv$ where $u = t_1u_1 + \ldots + t_{m-1}u_{m-1} + t_{m+1}u_{m+1} + \ldots + t_ru_r$, $t_m = t$ and $u_m = v$, the function defined by

$\beta(t_1, \ldots, t_r) = (\hat{f}^{(k)}y)\ (t_1u_1 + \ldots + t_ru_r)$ is a polynomial of degree $\leq k$ in t_m, if $|t_m| \leq 1$, for every $m = 1,2,\ldots,r$. Hence $\beta(t_1, \ldots, t_r)$ is a polynomial in all t's together of degree $\leq rk$.

Since β is homogeneous of exactly degree k that is, $\beta(st)^k = s^k\beta(t^k)$ for $s \geq 0$ every non-zero term in the expression of β has degree exactly k and so

$$\beta(t_1,\ldots,t_r) = \sum_{i_1+\ldots+i_k = k} \alpha_{i_1,\ldots,i_k} \left(\prod_{j=1}^{k} t_{i_j}\right),$$

and the coefficients can be choosen so that whenever $\mu : [1,k] \to [1,k]$ is a permutation then

$$a_{i_{\mu 1}, i_{\mu 2}, \ldots, i_{\mu k}} = a_{i_1,\ldots,i_2,\ldots,i_k}$$

We define a function

$$\alpha(t^1,t^2,\ldots,t^k) = \sum_{i_1+\ldots+i_k = k} a_{i_1,i_2,\ldots,i_k} \cdot t^1_{i_1} t^2_{i_2} \ldots t^k_{i_k}$$

where each $t^j = (t^j_1, t^j_2,\ldots,t^j_k)$ is a sequence of r number.

Since $u_1,\ldots,u_r$ is a basis for a finite dimensional subspace of E, the correspondence $t \to t_1u_1+\ldots+t_ru_r$, determines $(f^{(k)}y)$, corresponding naturally to α on the finite dimensional subspace such that $(\hat{f}^{(k)}y)u = (f^{(k)}y)u^k$. But $(f^{(k)}y)$ can then be obtained from $(\hat{f}^{(k)}y)$ by the ~ operation (cf Remark. 3.1 and expression (3.3)) $(f^{(k)}y) = (\tilde{\hat{f}}^{(k)}y)$ so that $(f^{(k)}y)$ is determined on all E and has a bound that is determined by the bound on $(\hat{f}^{(k)}y)$ (cf. (3.4)).

THEOREM 3.2 Let $f : A \to R$, A being an open subset of a Banach space E, be n-convex locally bounded and let one-sided (n-1) discrete differential $\hat{f}^{(n-1)}$ (cf. Lemma 3.3) be continuous at x. Then there is a bounded (n-1) linear function $(f^{(n-1)}x)$ such that $(\hat{f}^{(n-1)}x)u = (f^{(n-1)}x)u^{n-1}$ and

(i) $f^{(k)}$, $k = 0,1,2,\ldots,n-2$, has a strong Taylor series expansion about x of the form

$$f^{(k)}(x+su) = \sum_{j=k}^{n-1} (1/(j-k)!).(f^{(j)}x)\,(su)^{j-k} + r,$$

where $r/s^{n-1-k} \to 0$ uniformly for $||u|| \le 1$.

(ii) $(f^{(n-1)}x)$ is the Frechet differential of $f^{(n-2)}$ at x.

LEMMA 3.5 ([4], Theorem 3.2 p.269). Let E and F be two Banach spaces, A be an open subset of E. Let $f : A \to F$ and f have a weak n-Taylor series expansion (3.1) about every point of some neighbourhood N of a point $x \in A$. If y and u are vectors such that the line segment $\{y+tu; 0 \le t \le 1\}$ is in N and P_n is continuous at x. Then for $0 \le k \le n, P_k$ has an expansion about y of the form

$$P_k\,(y+su) = \sum_{j=k}^{n} (1/(j-k)!).\ (P_j(y))u^{j-k} + r_k(s,u)$$

where the remainder $r_k(s,u)$ has the property that for every $\in > 0$. There exists a $\delta > 0$ such that whenever

$$||y-x|| < \delta/2, \quad |s| < \delta/4 \quad ||u|| \le 1 \text{ then}$$

$$||r_k(s,u)|| < \in |s|^{n-k}$$

In particular P_k has a strong (n-k)-Taylor series expansion about x.

The proof is given in [4].

PROOF OF THEOREM 3.2

(1) If we define

$$[P_\eta(x,u)]t = \sum_{k=0}^{n-1} \left(\prod_{j=0}^{k-1} (t-j\eta)\right).\ [\Delta_k h(x,v)]\,(0,\eta,\dots,k\eta)$$

Then an argument similar to that used to prove Theorem 3.1 shows that $f(x+tu) - [P_\eta(x,u)]t$

$$=\left(\prod_{j=1}^{m-1} (t-j\eta)\right).[\Delta_{n-1}h(x,u)](\eta,\ldots,(n-1)\eta,t)-[\Delta_{n-1}h(x,u)](0,\eta,..,(n-1)\eta)$$

By monotonicity of $[\Delta_{n-1}h(x,u)]$ (cf. Lemma 3.1) both values of $[\Delta_{n-1}h(x,u)]$ are between $(\hat{f}^{(n-1)}x)u$ and $(\hat{f}^{(n-1)}(x+tu))u$ and continuity of $\hat{f}^{(n-1)}$ at x implies that $[f(x+tu) - [P_\eta(x,u)]t]/t^{n-1} \to 0$, as $t \to 0$ uniformly for $||u|| \leq 1$.

An argument similar to the proceeding shows that $(\hat{f}^{(n-1)}x)$ is determined by a bounded (n-1) linear function $(f^{(n-1)}x)$. By making $\eta \to 0$ and using Lemma 3.3 gives

$$\left|(f(x+tu) - \sum_{k=0}^{n-1} (t^k/k!)\,(f^{(k)}y)u^k)\right|/t^{n-1} \to 0 \text{ as } t \to 0 \text{ uniformly for}$$

$||u|| \leq 1$.

Thus it gives that f has a strong Taylor series expansion. Replacing n by n - 1 in Lemma 3.5 we get the expansion $f^{(k)}(x+su)$ and it proves (i). Then (ii) is an easy consequence of the expansion of $f^{(n-2)}$ in (i).

THEOREM 3.3 Let $f : A \to R$, A being an open set of a Banach space E, be n-convex and locally bounded. If for some $x \in A$, there is a sequence $\alpha_0,\alpha_1,..\alpha_n$, where α_k is bounded n-linear function such that f has a strong Taylor series expansion :

$$f(x+u) = \sum_{k=0}^{n} (1/k!)\,\alpha_k u^k + r(x+u), \quad \text{where } \lim_{u\to 0} \frac{r(x+u)}{||u||^n} = 0.$$

Then $\hat{f}^{(n-1)}$ is continuous at x and has a Fréchet differential at x with values $[(\hat{f}^{(n-1)}x)'u]v = \alpha_n(u,v^{n-1})$. In other words nth Fréchet differential of f exists.

PROOF. It is sufficient to show that

$(\hat{f}^{(n-1)}y)v = \alpha_{n-1}\, v^{n-1} + \alpha_n((y-x),\, v^{n-1}) + R$, where $\lim_{y\to x} R/||y-x|| = 0$,

uniformly for $|v| \le 1$.

To estimate $\hat{f}^{(n-1)}$ for any $\epsilon > 0$, choose $\delta > 0$ such that if $||y-x|| < \delta/2$, $||ry|| \le \epsilon\, ||y-x||^n$.

Then for any y such that $||y - x|| < \delta/2$ and $v \in E$, such that $||v|| \le 1$, let

$$Py = \sum_{k=0}^{n} (1/k!)\,(\alpha_k(y-x)^k)$$

Then

$$f(y+tv) = P(y+tv) + r(y+tv)$$

Let $[h(y,v)]t = f(y+tv)$, $[\bar{P}(y,v)]t = P(y+tv)$ and $[\bar{r}(y,v)]t = r(y+tv)$.

Since $\bar{r}(y,v) = h(y,v) - \bar{P}(y,v)$

$$[\Delta_{n-1}\bar{r}(y,v)]t = [\Delta_{n-1}h(y,v)]t - [\Delta_{n-1}\bar{P}(y,v)]t$$

For any $T > 0$. let $t = (0, T/(n-1), 2T/(n-1), \ldots, T)$. Then

$$[\Delta_{n-1}\bar{r}(y,v)]t \le (\epsilon . n(n-1)^{n-1})\ \frac{(||y-x|| + T)^n}{T^{n-1}},$$

provided $||v|| \le 1$, and $|T| < \delta/2$.

Now letting $A = n(n-1)^{n-1}$ and $T = \epsilon^{1/n}\ ||y-x||$ we have

$$|\ [\Delta_{n-1}h(y,v)]t - [\Delta_{n-1}\ \bar{P}(y,v)]\ t| \le A\,\epsilon^{1/n}(1+\epsilon^{1/n})^n\ ||y-x||$$

and similarly

$$|[\Delta_{n-1}h(y,v)](-t) - [\Delta_{n-1}\bar{P}(y,v)](-t)| \le A\epsilon^{1/n}(1+\epsilon^{1/n})^n \,||y-x||$$

But P is an n th degree polynomial, so P is n-convex and we may assume $[\Delta_n P(y,v)]\, t \ge 0$. Then

$$\left(\frac{1}{(n-1)!}\right) P^{(n-1)}(y-T,v)\, v^{n-1} - A\epsilon^{1/n}(1+\epsilon^{1/n})^n \,||y-x||$$

$$\le [\Delta_{n-1}h(y,v)]t$$

$$\le (1/(n-1)!)P^{(n-1)}(y+Tv)v^{n-1} + A\epsilon^{1/n}(1+\epsilon^{1/n})^n \,||y-x||$$

and $[\Delta_{n-1}h(y,v)]\,(-t)$ lies between the same bounds. By n-convexity $(\hat{f}^{(n-1)}y)v$ is again with in the same bounds. By elementary calculations

$$(P^{(n-1)}(y+tv))v^{n-1} = \alpha_{n-1}v^{n-1} + \alpha_n(y-x,\, v^{n-1}) + t\alpha_n v^n,$$

so that

$$-A\epsilon^{1/n}(1+\epsilon^{1/n})^n \,||y-x|| - (T\alpha_n v^n)/(n-1)!$$

$$\le (1/(n-1)!)[(\hat{f}^{(n-1)}y)v - (\alpha_{n-1}v^{n-1}) - (\alpha_n(y-x,\, v^{n-1})]$$

$$\le A\epsilon^{1/n}(1+\epsilon^{1/n})^n \,||y-x|| + (T\alpha_n v^n)/(n-1)!.$$

It follows by definition of T, that

$$(\hat{f}^{(n-1)}y) = \alpha_{n-1}v^{n-1} + \alpha_n(y-x,\, v^{n-1}) + R$$

where $R/||y-x|| \to 0$ as $y \to x$ uniformly for $||v|| \le 1$.

In view of Theorem (3.2) final result follows.

The author is highly obliged to Professor E.B. Leach, of Case Western Reserve University, Cleveland, Ohio for his valuable help and advise.

REFERENCES

[1] J.M. Ash. A characterization of the Peano derivatives Trans. Amer. Math. Soc. 149 (1970) 489-501.

[2] P.L. Butzer and W. Kozakiewez. On Riemann derivatives of integrated functions. Cand. J. Math. 6 (1954) 572-581.

[3] S.Dayal, Local representation of function on normal linear spaces, Ph.D. Thesis, Case Western Reserve University, Cleveland Ohio, 1972.

[4] S.Dayal, A converse of Taylor's Theorem for functions on banach spaces, Proc. Amer. Math. Soc. Vol. 65, No. 2, Aug. 1977, 265-273.

[5] S.Dayal, K-discrete differential of certain operators on Banach spaces, Proc. Amer. Math. Soc. Vol. 83, No. 1 1981, 77-82.

[6] J. Dieudonné, Foundations of Modern Analysis, Academic Press, New York 1960 MR 12 # 110074.

[7] T.M. Flett, Differential Analysis, Cambridge University Press, Cambridge, 1980.

[8] R. Ger, n-convex functions in linear spaces, Aeqationes Math 10 (1974), 172-176.

[9] R. Ger, Convex functions of higher orders in Euclidean spaces. Ann. Polon. Math. 25 (1972) 293-302.

[10] Hilderbrand, F.B. Introduction to Numerical Analysis, McGraw-Hill, New York, 1956, MR 17 # 788.

[11] J. Kopec and J. Musielak, On the estimation of the norm of the n-linear symmetric operators, Studia Math. 15 (1955), 29-30, MR 17, # 5 12.

[12] E.B. Leach, Differential calculus of sub-convex functionals. Unpublished note, 1968.

[13] E.B. Leach, and Whitefield, J.H.M., Differentiable functions and rough norms on Banach spaces, Proc. Amer. Math. Soc. Vol. 33, Number 1, 1972, 120-126.

[14] J. Kolomy, On the Differentiability of Mapping and Convex Functionals (Commt. Math. Univ. Carolinae 84 (1967) 735-751).

[15] M.Z. Nashed Differentiability and related properties of non-linear operators. Some aspects of the role of differentials in non-linear analysis, Non-linear Functional. Anal. and Appl. (Proc. Advanced Sem. Math. Res. Centre, Univ. of Wisconsin, Madison 1970) Academic Press, New York 1971. pp. 103-309.

[16] A.E. Taylor, Addition in the theory of polynomials in normed linear spaces, Tôhoku Math. J. 44 (1938), 302-318.

[17] C.E. Weil, On approximation and Peano derivatives, Proc. Amer. Math. Soc. 20 (1969), 487-490.

Department of Mathematics,
Maharshi Dayanand University,
Rohtak-124001 (INDIA).

ROLE OF JAMES LIKE SPACES IN MULTIPLICATIVE LINEAR FUNCTIONALS ON OPERATOR ALGEBRAS

SHASHI KIRAN and AJIT IQBAL SINGH

DEDICATED TO THE MEMORY OF U.N. SINGH

1. INTRODUCTION

(1.1) Since the only finite dimensional Banach Space Y for which the operator algebra B(Y) has a non-zero multiplicative linear functional is that of dimension one it is difficult to imagine that a non-zero multiplicative linear functional exists on the operator algebra B(Y) of an infinite dimensional Banach Space Y. (Henceforth we shall use the abbrevation mlf for a non-zero multiplicative linear functional on a Banach algebra.)

(i)a) The first such example came in 1970 when Mityagin and Edelstein ([7], § 2.2 p225) obtained an mlf on B(J) where J is the classical James Space, the first non-reflexive space isometrically isomorphic with its bidual J^{**} and of codimension one in J^{**} (i.e. $J^{**}/J = R$) ([3], [4]).

(i)b) Mityagin and Edelstein [7] further showed that for n ≥ 2 the operator algebras of the spaces J^n (which are quasi-reflexive of order n i.e. have co-dimension n in their biduals) do not possess any mlf's.

(ii) Mityagin and Edelstein, in the same paper ([7], § 2.9, p 230) also gave an mlf on the Banach algebra $B(C(\overline{\downarrow w_1}))$ where $C(\overline{\downarrow W_1})$ is the space of all continuous scalar valued functions on the set of ordinals not exceeeding the first uncountable ordinal W_1 with its usual order topology, equipped with the supremum norm.

(iii) Simultaneously and independently, A.Wilansky ([11], Thm1, Thm2) gave, for a non-reflexive space Y, a method of obtaining an mlf on a closed subalgebra of B(Y). In the case of J and $C(\overline{\downarrow W_1})$ it is the whole operator algebra and moreover the mlf obtained is also the same.

(iv) Another Banach Space Y such that B(Y) admits an mlf is given in Shelah and Steprans [9] (see also [8]). They constructed a non-separable space Y on which every operator has the form $S+\rho I$ where ρ is a scalar and S is an operator with a separable range. B(Y) for this Y admits an mlf.

(1.2) In the example given in (1.1) above the underlying space was non-reflexive. Moreover examples of B(Y) admitting more than one mlf were not known till Mankiewicz ([6], Thm 1.1, Cor 6.7) constructed a separable super reflexive Banach Space Y such that the Banach algebra B(Y) admits a continuous homomorphism h onto l_N^∞ or equivalently $C(\beta N)$-the Banach algebra of all scalar valued continuous functions on the Stone-Čech compactification of the set N of natural numbers, equipped with the supremum norm.

(1.3) In this paper we study the problem of existence and nature of mlf's on B(Y) with $Y=J(X_n)$ [1] or J(X)[12] and thus enlarge the class of operator algebras admitting mlf's. In particular we show that when X is reflexive, the cardinality of mlf's on B(J(X)) is at least equal to that on B(X).

2. PRELIMINARIES

(2.1) NOTATION: In this paper we consider real Banach Spaces only. A Banach Space Y will often be identified with its canonical image in its bidual Y^{**}. For a closed subspace Y_1 of a Banach Space Y, Y/Y_1 is the quotient Banach Space and an element $y+Y_1$ will usually be abbreviated to [y]. $\cong$ stands for isometric isomorphism of Banach Spaces. The term homomorphism stands for a linear multiplicative map from one algebra to another. For an operator T on a Banach Space Y, T^{**} stands for the double adjoint of T.
The identity operator on Y will be denoted by I_y.

(2.2) The Operation Q: (i) For a Banach Space Y and an operator T on Y, Q(T) is the operator on Y^{**}/Y defined by $Q(T)(x^{**}+Y)=T^{**}x^{**}+Y$ for $x^{**} \in Y^{**}$, or in our notation,

$Q(T)[x^{**}]=[T^{**}x^{**}]$. $Q(T)$ is the same as $H(T)$ in ([12], §2,p 963). Then as noted in ([13], Thm 1.1) Q is a homomorphism on the algebra $B(Y)$ into $B(Y^{**}/Y)$ which takes I_Y to $I_{Y^{**}/Y}$. This also shows that Q takes an invertible operator on Y to an invertible operator on Y^{**}/Y.

(ii) For a closed subspace W of Y^{**} with $Y \subset W$ we let $B_W(Y)=(T \in B(Y):T^{**}W \subset W)$. Then $B_W(Y)$ is a closed subalgebra of $B(Y)$ and we may define $Q_W:B_W(Y) \to B(Y^{**}/W)$ via $Q_W(T)(x^{**}+W)=T^{**}(X^{**})+W$, $x^{**} \in Y^{**}$.

Then Q_W is a homomorphism, $Q_W(I_y)=I_{y^{**}/W}$ and Q_W takes an invertible element of $B_W(Y)$ to an invertible element of $B(Y^{**}/W)$.

(2.3) James Sum of Banach Spaces: (i) As a particular case of ([1],§1), let (X_n) be an increasing sequence of closed subspaces of a Banach Space Z such that $\cup X_n$ is dense in Z. For notational convenience, we set $X_o = \{0\}$ and $x_o=0$. For $x=(x_n)$, $x_n \in X_n$ for each n, and for a finite increasing sequence $P=\{p_1,p_2, \dots p_k\}$ of non-negative integers let

$$||x||_P^2 = \sum_{i=1}^{k-1} ||x_{p_i} - x_{p_{i+1}}||^2+|| x_{p_k}||^2$$

and $\sqrt{2}||x||_J= \sup ||x||_P$,

where the supremum is taken over all such finite increasing sequences $p=\{p_1,p_2,\dots p_k\}$ of all non negative integers. The James Sum $J(X_n)$ of Banach Spaces (X_n) is defined as $J(X_n) = \{x=\{x_n\},\ x_n \in X_n$ for all n, $||x||_J < \infty$ and $\lim x_n=0\}$ which equipped with the norm $||.||_J$ is a Banach Space. ([1], §1, Remark 3) and proof of ([1],Thm 1.1) say that each X_n is embedded isometrically in $J(X_n)$ and (X_n) is a monotone shrinking decomposition for $J(X_n)$ if we identify for each n, X_n with $G_n =\prod X_j^n$ where $X_n^n = X_n$ and $X_j^n = \{0\}$ for $j \neq n$. We shall often use this identification.

(ii) $J^{LIM}(X_n) = \{x=(x_n): x_n \in X_n$ for all n, $||x||_J < \infty \}$ equipped with the norm $||.||_J$ is also a Banach Space. It turns out that for each $x = (x_n) \in J^{LIM}(X_n)$, $\lim_{n\to\infty} x_n$ exists in Z.

(iii) If i_n is the inclusion map from X_n into Z, i_n^{**} may be looked upon as inclusion map from X_n^{**} into Z^{**}. Thus the sequence (X_n^{**}) is an increasing sequence of Banach Spaces. So we can construct $J(X_n^{**})$ and $J^{LIM}(X_n^{**})$. Z_1 denotes the closure of $U_n X_n^{**}$ in Z^{**}.

As a consequence of ([1], Prelim.) (See also [10], §III.15 : [5], § 1), $J(X_n)^{**}$ is isometrically isomorphic to $J^{LIM}(X_n^{**})$.

(iv) The quotient $J(X_n)^{**} / J(X_n^{**}) \cong Z_1$ and the quotient map λ on $J(X_n)^{**}$

or $J^{LIM}(X_n^{**})$ is given by $\lambda((x_n)) = \lim x_n$ for $(x_n) \in J^{LIM}(X_n^{**})$.
(v) If each X_n is reflexive, $J(X_n)^{**} \cong J^{LIM}(X_n)$ and $J(X_n)^{**}/J(X_n) \cong Z$. ,

(vi) If $X_n = X$ for each n, as in [15], $J^{LIM}(X^{**}) \cong J(X)^{**} \cong J(X^{**}) \oplus X^{**}$.

(2.4) OPERATOR ALGEBRA $B(J(X_n))$ (i) As in [5], § 2.5) Let Inv $B(J^{LIM}(X_n)) = \{T \subset B(J^{LIM}(X_n)) : T X_n \subset X_n$ for all n } and Inv $B(J(X_n)) = \{T \in B(J(X_n)) : T X_n \subset X_n$ for all n) For a $T \in$ Inv $B(J(X_n))$ we put $T_n =$ T/X_n. Then $(T_n) \in \prod_{n=1}^{\infty} B(X_n)$ and T is the unique member of $B(J(X_n))$ satisfying $T/X_n = T_n$ Simple computations as in ([10], § III.15) show that $T^* f = (T_n^* f_n)$ and $T^{**}(F) = (T_n^{**} F_n)$ for $f = (f_n) \in J(X_n)^*$ anf $F = (F_n) \in J^{LIM}(X_n^{**})$. Since for $(F_n) \in J(X_n^{**})$ we have for $n \in N$, $||T_n^{**} F_n|| \le ||T||\ ||F_n|| \to 0$ we conclude that $T^{**} F \in J(X_n^{**})$ whenever $F \in J(X_n^{**})$ i.e. $T^{**}(J(X_n^{**})) \subset J(X_n^{**})$. Hence in the notation of (2.2) (ii)

$$\text{Inv } B(J(X_n)) \subset B_{J(X_n^{**})}(J(X_n))$$

(ii) Let $B_o(Z) = \{T \in B(Z) : TX_n \subset X_n$ for all n $\}$ and $B_o^{**}(Z_1)=(T^{**}/Z_1 : T \in B_o(Z)\}$. It follows from ([5], Thm 2.6(i)) or by direct computation that $T \in B_o(Z)$ induces an operator $S_T \in \text{Inv } B(J(X_n))$, defined by: For $x = (x_n) \in J(X_n)$, $S_T((x_n)) = (Tx_n) = (T_n x_n)$ where $T_n = T/X_n$ for each n.

(iii) If $X_n = X$ for each n, S_T is the operator J(T) as in ([12], § 2, p 962 $J(T)((X_n)) = (Tx_n$.

3. MLF's On Operator Algebras

We are now ready to give mlf's on $B_{J(X_n^{**})}(J(X_n))$ in case $B(Z_1)$ possesses mlf's. We shall abbreviate $B_{J}(X_n^{**})(J(X_n))$ and $Q_J(X_n^{**})$ to $B_1(J(X_n))$ and Q_1 respectively in this section.

(3.1) **THEOREM:** If $B(Z_1)$ has an mlf then $B_1(J(X_n))$ has an mlf and therefore so does every closed subalgebnra A of $B_1(J(X_n))$ containing $I_{J(X_n)}$.

PROOF. By (2.2) (ii) the operator Q_1: $B_1(J(X_n)) \dashrightarrow B(Z_1)$ is a homomorphism taking $I_{J(X_n)}$ to I_{Z_1}. So for an mlf ϕ on $B(Z_1)$, $\phi o Q_1$ is an mlf on $B_1(J(X_n))$.

Since $B_1(J(X_n))$ coincides with the whole operator algebra and Z_1 coincides with Z when each X_n is reflexive we have the following special case.

(3.2) **THEOREM:** If each X_n is reflexive and if B(Z) has an mlf, then $B(J(X_n))$ has an mlf.

We now come to the cardinality of mlf's on $B_1(J(X_n))$, $B(J(X_n))$ and $B_J(X)$.

(3.3) **PROPOSITION:** Let ϕ,ϕ' be two mlf's on $B(Z_1)$ such that $\phi\big|_{B_o^{**}(Z_1)} \neq \phi'\big|_{B_o^{**}(Z_1)}$. Then $\phi o\, Q_1\big|_{\text{Inv}B(J(X_n))} \neq \phi' o Q_1\big|_{\text{Inv}B(J(X_n))}$

PROOF: It is enough to show that $Q_1(\mathrm{InvB}(J(X_n))$ contains $B_o^{**}(Z_1)$. Let $T \in B_o(Z)$. Then S_T as defined in (2.4) (ii) is in $\mathrm{InvB}(J(X_n))$. We show that $Q_1(S_T) = T^{**}/Z_1$.

For $z \in Z_1$ and a sequence $(F_n) \in J(X_n^{**})$ such that $\lambda((F_n)) = [(F_n)] = \lim F_n = z$ where λ is the quotient map as in (2.3) (iv) we have $[(F_n)] = z$. Now $Q_1(S_T)(z) = \lim T^{**} F_n = T^{**} \lim F_n = T^{**} z = (T^{**}/Z_1)(z)$

This completes the proof.

(3.4) **THEOREM:** Let α be the cardinal number of the set of mlf's which are distinct on $B_o^{**}(Z_1)$.

(i) For any closed subalgebra A of $B(J(X_n))$ lying between Inv $B(J(X_n))$ and $B_1(J(X_n))$ the cardinal number of the set of mlf's on A is greater than or equal to α.

(ii) In particular if each X_n is reflexive then $B(J(X_n))$ has at least as many mlf's as $B_o(Z)$ does.

(iii) Finally if X is a reflexive Banach Space then B(J(X)] has at least as many mlf's as B(X) does.

(3.5) **REMARKS** (i) The last part of Theorem 3.4 may be viewed as generalisation, of Mityagin and Edelstein's result given in (1.1) above, from R to a reflexive Space X.

(ii) Mityagin and Edelstein's method can also be directly applied to the James like spaces $J(x_i)$ given by Lohman and Casazza [2] for a Banach Space with a symmetric boundedly complete p-Hilbertian $(1<p<\infty)$ basis (x_i) since $J(x_i)^{**} = J(x_i) \oplus R$ as is the case with J.

(iii) We consider the case when each X_j is finite dimensional so that $Z_1 = Z$ and $B_o^{**}(Z_1) = B_o(Z)$ denoting $J(X\ )$ by $\hat{Z}$ and for $j \in N$,

the space $\{(X_n) \in J(X_n) : X_n = 0 \text{ for } n > j\}$ by $\hat{X}_j$, $\hat{X}_j$'s form an increasing sequence of finite dimensional subspaces of $J(X_n)$ with

$\overline{U\hat{X}_j}=J(X_n) = \hat{Z}$ so we may construct $J(\hat{X}_n)$ and denote it by (Z). Thus we may inductively define a sequence $Z^{(n)}$ of separable Banach Spaces by taking $Z(1) = Z$, $Z^{(n+1)}=(z^{(n)})^{\wedge}$.

a) Theorem 3.1 thus gives us a sequence of Banach Spaces $Z^{(n)}$ such that each $B(Z^{(n)})$ has an mlf as soon as B(Z) does.

In particular we may take Z=J or $J(x_i)$ as in (ii) above and for $j \in N$, $X_j = \{(a_n) \in Z : a_n = o \text{ for } n > j\}$ We may also take Z to be the space constructed by Mankiewicz [6] referred to in (1.2) above. The example is very pathological and is based on the technique of random finite dimensional Banach Spaces. In the notation of ([6], § 4) we take

$$Z = (\overset{\infty}{\underset{k=1}{\oplus}} Y_{n_k}^{q_k})_{l^2} \text{ and for } j \in N,\ X_j = (\overset{j}{\underset{k=1}{\oplus}} Y_{n_k}^{q_k})_{l^2}$$

(b) We note that $B_o(\hat{Z})=B_o(J(X_n))$ contains Inv B $(J(X_n))$. Applying at each inductive step, this argument, viz,

$B_o Z^{(j+1)}= B_o(\hat{Z}^{(j)})=B_o(J(\hat{X}_n^{(j)}) \supset \text{Inv } B\ (J\ (\hat{X}_n^{(j)}))$ and theorem 3.4 (i) to $Z_o^{(j)}$ we have that for each j, the cardinal number of mlf's on $B(Z^{(j)})$ which are distinct on $B_o(Z^{(j)})$ is greater than or equal to α. So the sequence $(Z^{(n)})$ is such that for each n, cardinality of mlf's on $B_o(Z^{(n)})$ is greater than or eual to α.

(iv) We may apply theorem 3.4(iii) to the space given by Mankiewicz by taking $X_n = Y$ for each n and the thus constructed B(J(Y)) has at least as many mlf's as B(Y) i.e. 2^c many.

(v) We may make use of the space Y of Mankiewicz and our results in another way by taking as in (a) above,

$$Z = Y = (\overset{\infty}{\underset{k=1}{\oplus}} Y_{n_k}^{q_k})_{l^2} \quad \text{and} \quad X_j = (\overset{j}{\underset{k=1}{\oplus}} Y_{n_k}^{q_k})_{l^2}$$

for $j \in N$ and then constructing $J(X_n)$. Since the operator $\sum_{i=1}^{\infty} \alpha_i P_{N_i}$

([6],p 14, line14) in the proof of ontoness of the homomorphism h (see (1.2) above) is in fact in $B_o(Z)$ we have that $h(B_o(Z)) = l^{\infty}_N$ and thus distinct mlf's ψ and ψ' on $l^{\infty}_N \cong C(\beta N)$ give rise to mlf's $\phi = ho\psi$ and $\phi' = ho\psi$ and $ho\psi'$ on B(Z) which are distinct on $B_o(Z)$. As a conseuence the cardinality α of mlf's on B(Z) distinct on $B_o(Z)$ is 2^c.

Finally applying (iii) (b) above we can have a sequence $Z^{(n)}$ of separable Banach Spaces with the cardinal number of the sets of mlf's on $B(Z^{(n)})$ greater than or equal to 2^c for each n.

REFERENCES

[1] Bellenot, S.F. : The J-Sum of Banach Spaces, J. Funct. Anal. 48 (1982), 95-106.

[2] Casazza, P.G. and Lohman, R.H. : A general construction of spaces of the type of R.C.James, Canad. J. Math. 27 (1975), 1263-1270.

[3] James, R.C. : Bases and Reflexivity of Banach Spaces, Ann. of Math. 52 (1950), 518-527.

[4] ---------- : A nonreflexive Banach Space isometric with its second conjugate space, Proc. Nat. Acad. Sci. U.S.A. 37 (1951) 174-177.

[5] Kiran, Shashi and Singh A.I. : The J-Sum of Banach Algebras and some Applications, Yokohama Mathematical Journal 36 (1988) 1-20.

[6] Mankiewicz, P. : Superreflexive Banach Space X with L(X) admitting a homopmorphism onto the Banach Algebra C(βN), Israel J.Math. 65 No. 1 (1989) 1-16.

[7] Mityagin, B.S. and Edelstein I.S. : Homotopy type of linear groups of two classes of Banach Spaces,Functional Analysis and applications, 4 (3) (1970) 221-230 (English Translation).

[8] Shelah, S. : A Banach Space with a few Operators, Israel J. Math 30 (1978) 181-191.

[9] Shelah, S. and Steprans, J : A Banach Space on which there are few operators, Proc. Am. Math. Soc. 104 (I) (1989) 101-105.

[10]Singer, I : Bases in Banach Spaces II, Springer Verlag, 1981.

[11]Wilansky, A. : Subalgebras of B(X), Proc. Am. Math. Soc. 29 (1971), 355-360.

[12]W´ojtowicz, M. : Finitely nonreflexive Banach Spaces, Proc. Am. Math. Soc. 106 No. 4 (1989) 961-965.

[13]Yang, K.W. : The Reflexive dimension of an R-Space, Acta Math Hungar 35 (1980) 249-255.

SHASHI KIRAN
DEPARTMENT OF MATHEMATICS
MATA SUNDRI COLLEGE
UNIVERSITY OF DELHI
DELHI.

AJIT IQBAL SINGH
DEPARTMENT OF MATHEMATICS
PUNJAB UNIVERSITY
CHANDIGARH.

The Carleman-Fourier Transform And Its Applications

U.N. Singh

1. INTRODUCTION.

Carleman [1] generalised the concept of Fourier transform in the form of a pair of analytic functions one of which is holomorphic in the upper half plane and the other in the lower half plane. We have called this generalized Fourier transform 'Carleman-Fourier transform'. Following Carleman we have introduced in section 1 the notion of Carleman-Fourier transform which can be defined for a much wider class of functions than the L_p classes. Some lemmas have been proved in section 2 and in sections 3 and 4, we prove two theorems which extend two theorems of Hille and Tamarkin to a much wider class. In section 5 we extend an important classical theorem of Paley and Wiener.

The class of complex valued functions f, measurable on the whole real line, and Lebesgue-integrable in every finite interval will be designated C(k) if every f satisfies the condition

$$(1.1) \qquad \int_0^x |f(t)|dt = O(|x|^k), \quad \text{for } |x| \to \infty,$$

where k is a non-negative finite real number.

The class of functions f which, in addition to satisfying the condition (1.1) also satisfy the condition

$$(1.2) \qquad \int_0^x |f(t)|dt = O(|x|^k) \quad \text{for } x \to 0, \text{ with the same k as in (1.1),}$$

will be designated $C'(k)$. Thus $C'(k)$ $(k \neq 0)$ is a proper subset of $C(k)$. Since every function $f \in L_1(-\infty,\infty)$ satisfies both the conditions (1.1) and (1.2) at the same time with k=0, both the classes C(0) and C'(0) are identical with L_1. It can easily be seen by applying Holder's inequality that every $f \in L_p(-\infty,\infty)$, for $1 < p < \infty$ belongs to the class $C'(1/q)$, where

$q=p/(p-1)$. However, there are functions which do not belong to any L_p and belong to a $C'(k)$.

An analytic function ϕ, holomorphic in a half plane D of the complex plane, is said to belong to class (α,β), where α,β are real numbers such that $\alpha > 0$, $\beta > -1$, if

$$(1.3) \qquad \frac{(z-z_o)^{-\alpha}}{(z-z_o')^{\beta}\,\Gamma(\alpha)} \int_{z_o}^{z} (z-\xi)^{\alpha-1}\phi(\xi)d\xi \quad \text{is bounded in D,}$$

where z_o is an arbitrary point in the interior of D and z_o' is symmetric with respect to the straight line determining the half plane D. ϕ is said to belong to the class $(0,\beta)$, $\beta > -1$, if $(z-z_o')^{-\beta}\phi(z)$ is bounded in D.

Carleman has shown [1] that to every function $f \in C(k)$ there corresponds a couple of analytic functions $f_1(z)$ and $f_2(z)$, holomorphic for $Im(z) > 0$ and $Im(z) < 0$ respectively and such that

$$(1.4) \qquad \lim_{y\to+0} \int_{x'}^{x''} \{f_1(x+iy) - f_2(x-iy)\}dx = \int_{x'}^{x''} f(x)dx$$

uniformly in every closed interval $a \le x' \le x'' \le b$, where a and b are any finite real numbers. It is also true that

$$(1.5) \qquad \lim_{y\to+0} \{f_1(x+iy) - f_2(x-iy)\} = f(x) \qquad \text{a.e.}$$

In their regions of holomorphy the functions f_1 and f_2 belong to the finite class $(\gamma,k-1+\epsilon)$, $0 \le \epsilon < 1$, $[\epsilon = 0$ if k is non-integral$]$ and $1 < \gamma < 2$, and are unique upto an additional polynomial of degree $[k-1+\epsilon]$, where $[x]$ denotes the integral part of x. The couple of functions f_1 and f_2 is determined as follows.

Let $f \in C(k)$ and let $p = [k]$. Then the integral $\displaystyle\frac{p!}{2\pi i}\int_{-\infty}^{\infty}\frac{f(\xi)}{(\xi-z)^{p+1}}d\xi$ defines a holomorphic function in the complex plane cut along the real axis. If F denotes the primitive of the pth order of the function defined by the above integral, then

$$F^{(p)}(z) = \frac{p!}{2\pi i} \int_{-\infty}^{\infty} \frac{f(\xi)}{(\xi - z)^{p+1}} d\xi ,$$

and F(z) is determined upto an additional polynomial of degree (p-1). Then f_1 and f_2 are obtained as

$$f_1(z) = F(z) \quad \text{for } \mathrm{Im}(z) > 0,$$

$$f_2(z) = F(z) \quad \text{for } \mathrm{Im}(z) < 0.$$

Thus the function f_1 and f_2 are determined uniquely upto an additional polynomial. Hence onwards we shall simply write (f_1,f_2) for the couple of analytic functions $f_1(z)$ and $f_2(z)$.

Carleman [1] has also introduced the concept of `generalized Fourier transform´ for functions of class C(k). With every function $f \in C(k)$ he associates a couple of analytic functions $g_1(z)$ and $g_2(z)$, holomorphic respectively in the upper half plane and the lower half plane such that the couple (g_1,g_2) represents the Fourier transform g of f, whenever g is defined in the classical sense, in the same manner as the couple (f_1,f_2) represents the function f. Carleman calls the couple (g_1,g_2) the `generalized Fourier transform´ of the function $f \in C(k)$ or of the couple of functions (f_1,f_2) representing f. It is but appropriate that we call the couple (g_1,g_2) to be defined below, as the Carleman-Fourier transform of f or simply CF-transform of f. For $f \in C(k)$, the couple of CF-transform of f is defined as

$$(1.6) \qquad \frac{1}{\sqrt{2\pi}} \int_{-\infty}^{0} f(\xi)e^{-i\xi z}d\xi = g_1(z) \qquad \text{for } \mathrm{Im}(z) > 0,$$

$$(1.7) \qquad -\frac{1}{\sqrt{2\pi}} \int_{0}^{\infty} f(\xi)e^{-i\xi z}d\xi = g_2(z) \qquad \text{for } \mathrm{Im}(z) > 0.$$

Since $f \in C(k)$, the integrals in (1.6) and (1.7) define analytic functions $g_1(z)$ and $g_2(z)$ which are holomorphic in the upper half and lower half planes respectively.

To justify the above definition, let $f \in L_1(-\infty,\infty)$ and suppose that g is the Fourier transform of f. Now $g_1(x+iy)$ and $g_2(x-iy)$ as defined by (1.6) and (1.7) are holomorphic respectively in the upper half plane and the lower half plane and each one of them is continuous on the real axis. It follows easily that

$$\lim_{y \to +0} \{g_1(x+iy)-g_2(x-iy)\} = \frac{1}{\sqrt{2\pi}} \int_{-\infty}^{\infty} f(\xi)e^{-i\xi x}d\xi = g(x) \quad \text{p.p}$$

If $f \in L_p(-\infty,\infty)$, $1 < p \leq 2$, and g is its Fourier-Plancherel transform, then $g_1(z)$ and $g_2(z)$ as defined by (1.6) and (1.7) with $f \in L_p$, are holomorphic for $\operatorname{Im}(z) > 0$, and $\operatorname{Im}(z) < 0$, respectively. Also it can be shown that $g_1(z) \in H^q$ in the upper half plane, when H^q is the Hardy space and $q = \frac{p}{p-1}$. Similarly $g_2(z) \in H^q$ in the lower half plane. Consequently

$$\lim_{y \to +0} \{g_1(x+iy)-g_2(x-iy)\} = \underset{y \to +0}{\overset{q}{\text{l.i.m.}}} \{g_1(x+iy)-g_2(x-iy)\} \quad \text{p.p.}$$

It can further be shown that each of these limits is equal to $g(x)$ p.p., when $g(x)$ is the Fourier-Plancherel transform of f.

We have now associated with a function $f \in C(k)$ a couple of analytic functions (f_1,f_2) which represents the function f and a couple of analytic functions (g_1,g_2) which represents the Fourier transform of f. The function f was involved directly in defining the two couples. The question arises whether it is possible to define CF-transform of a couple (f_1,f_2) without involving f and also if CF-transform (g_1,g_2) of (f_1,f_2) has been defined whether we can determine the couple (f_1,f_2) from (g_1,g_2). Following Carleman we proceed to define now the CF-transform of a couple of analytic functions, more general than the couple of analytic functions representing a function $f \in C(k)$.

Let (f_1,f_2) be a couple of analytic functions holomorphic respectively in the upper half plane and the lower half plane satisfying the following condition :

$$|f_1(re^{i\theta})| < A(\theta_o)(r^{\alpha}+1/r^{\beta}) \qquad \text{for } \theta_o < \theta < \pi-\theta_o,$$

(1.9)

$$|f_2(re^{i\theta})| < A(\theta_o)(r^{\alpha}+1/r^{\beta}) \qquad \text{for } -\pi+\theta_o < \theta < -\theta_o$$

for every θ_o such that $0 < \theta_o < \pi/2$ where $\alpha,\beta \geq 0$, and $z = re^{i\theta}$.

As we will see later, the couple (f_1,f_2) associated with $f \in C(k)$ satisfies an inequality of the type (1.9).

Let L and L′ be two rays (half lines) issuing forth from the origin and lying respectively in the upper half plane and the lower half plane of the complex plane. Consider the integrals

$$(1.10) \qquad G(z) = \frac{1}{\sqrt{2\pi}} \int_L f_1(u)e^{-iuz}du,$$

$$(1.11) \qquad H(z) = \frac{1}{\sqrt{2\pi}} \int_{L'} f_2(u)e^{-iuz}du,$$

If $\beta < 1$ in (1.9), and the rays L and L′ make with the positive real axis angles ϕ and $-\phi'$ respectively, $\phi,\phi' > 0$, then the integrals (1.10) and (1.11) define analytic functions G(z) and H(z) holomorphic respectively in the half planes $-\pi-\phi < \theta < -\phi$ and $-\pi+\phi' < \theta < \phi'$, where $z = re^{-i\theta}$ and $0<r<\infty$.

It is easily seen that the two functions G(z), given by the integral (1.10) corresponding to two positions of the ray L in the upper half plane, i.e. corresponding to two different values of ϕ, will be equal to each other in the region common to the two regions of holomorphy of G. Since the ray L can be rotated freely about the region in the upper half plane, it follows from the principle of analytic continuation that the function G is holomorphic in the finite part of the complex plane cut along the positive real axis, i.e. in the region $0 < \theta < 2\pi$, $0 < r < \infty$. It can be shown in the same manner that the function H is holomorphic in the finite part of the complex plane cut along the negative real axis, i.e. in the region $-\pi < \theta < \pi$, $0 < r < \infty$.

If $\beta \geq 1$ in (1.9), set $m = [\beta]$. The functions G and H are determined in this case from their derivatives of the mth order defined respectively by the following integrals :

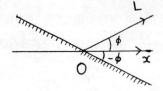

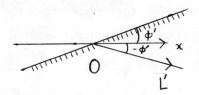

(1.12) $$G^{(m)}(z) = \frac{(\ i\ )^m}{\sqrt{2\pi}} \int_L u^m f_1(u) e^{-iuz} du,$$

and

(1.13) $$H^{(m)}(z) = \frac{(-i)^m}{\sqrt{2\pi}} \int_{L'} u^m f_2(u) e^{-iuz} du,$$

where the rays L and L' are the same as defined in the preceding paragraph. Following the same analysis as in the case $0 \le \beta < 1$, it can be shown that the function $G^{(m)}(z)$ is holomorphic in the finite part of the complex plane cut along the positive real axis, i.e. in the region $0 < \theta < 2\pi$, $0 < r < \infty$, and H(z) is holomorphic in the finite part of the complex plane cut along the negative real axis i.e. in the region $-\pi < \theta < \pi$, $0 < r < \infty$. Each of the functions G(z) and H(z) is determined uniquely upto an additional polynomial of degree (m-1).

Having done this Carleman has defined the couple (g_1,g_2) the 'generalized Fourier transform' of the couple (f_1,f_2) by the following relations :

(1.14)
$$g_1(z) = H(z)-G(z) \qquad \text{for } Im(z) > 0,$$
$$g_2(z) = H(z)-G(z) \qquad \text{for } Im(z) < 0.$$

Since each of the functions G and H is determined uniquely upto a polynomial of degree (m-1), it follows that the couple (g_1,g_2) is also determined uniquely upto a polynomial of degree (m-1). This really means that the couples (g_1,g_2) form a class, any two couples of which class differ between themselves at the most by a polynomial of degree (m-1). The couple (g_1,g_2), thus defined will be called, as stated earlier, the Carleman-Fourier transform or simply CF-transform of the couple (f_1,f_2).

The formula (1.14) represents a linear transformation of the couple (f_1,f_2) into another couple (g_1,g_2), called the CF-transform of (f_1,f_2), denoted symbolically as

$$(1.15) \qquad g = S(f).$$

Denoting by T(g) another transformation which transforms the couple (g_1,g_2) into the couple of analytic functions $\bar{g}_2(\bar{z})$ and $\bar{g}_1(\bar{z})$, which are holomorphic respectively for $\mathrm{Im}(z) > 0$ and $\mathrm{Im}(z) < 0$, we state the important Carleman-Fourier theorem as follows :

$$(1.16) \qquad TSTS(f) = f, \text{ upto a polynomial,}$$

where the additional polynomial is of the type $\sum_{r=0}^{(m'-1)} a_r z^r$

if in (1.9) $\beta < 1$, $m' = [\alpha+1]$, and of the type $\sum_{r=0}^{(m'-1)} a_r z^{r-m}$

if in (1.9) $\beta \geq 1$, $m = [\beta]$, $m' = [\alpha+m+1]$.

It may be observed here that :

If $f \in C'(k)$, $(0 < k < 1)$, and in particular if $f \in L_p$, $(1 < p < \infty)$, then $TSTS(f) = f$ exactly.

It can also be proved that for $f \in C(k)$, $0 \leq k < \infty$, the two definitions of (g_1,g_2), the CF-transform of f, one given by (1.6) and (1.7) and the other given by (1.14), are equivalent.

2. If F(z) be an analytic function which is holomorphic for $\mathrm{Im}(z) > 0$, $z = x+iy$, and satisfies the condition :

$$(2.1) \qquad \lim_{y \to +0} F(x+iy) = f(x) \text{ almost everywhere on the real axis, it will}$$

be said that f is the limit function of f.

If

$$(2.2) \qquad F(z) = \frac{1}{2\pi i} \int_{-\infty}^{\infty} \frac{f(\xi)}{\xi - z} d\xi \qquad \text{for } \mathrm{Im}(z) > 0,$$

F is said to be represented by the Cauchy integral of f. We propose to study conditions under which a function $f \in L_p$, $(1 \le p < \infty)$, or more generally $f \in C(k)$ can be represented by its Cauchy integral. Before we state our theorems and prove them, it is necessary to first prove some auxiliary results which will be needed in the study of our problem.

LEMMA 1. Let (f_1, f_2) be the function couple and (g_1, g_2) the CF-transform couple of a function $f \in C'(k)$, $0 < k < 1$. Then for $z = re^{i\theta}$, $0 < \theta_0 < \pi/2$,

$$(2.3)\quad \begin{aligned} |f_1(re^{i\theta})| &< A(\theta_0)\left(\frac{1}{r^{1-k}}\right), \text{ for } \theta_0 < \theta < \pi-\theta_0, \\ |f_2(re^{i\theta})| &< A(\theta_0)\left(\frac{1}{r^{1-k}}\right), \text{ for } \pi+\theta_0 < \theta < -\theta_0, \end{aligned}$$

for every θ_0 ;

and

$$(2.4)\quad \begin{aligned} |g_1(re^{i\theta})| &< A_1(\theta_0)\left(\frac{1}{r^{k}}\right), \text{ for } \theta_0 < \theta < \pi-\theta_0, \\ |g_2(re^{i\theta})| &< A_1(\theta_0)\left(\frac{1}{r^{k}}\right), \text{ for } -\pi+\theta_0 < \theta < -\theta_0, \end{aligned}$$

for every θ_0.

PROOF. Take the inequalities (2.3) first and here also we may just prove the inequality for f_1, the proof for f_2 being similar. Let $F(\xi) = \int_0^{\xi} f(t)dt$. Then for every real ξ $|F(\xi)| \le \int_0^{\xi} |f(t)|dt < \lambda|\xi|^k$, where λ is a constant since $f \in C'(k)$, and if $z = x + iy = re^{i\theta}$ is situated in the region $\theta_0 < \theta < \pi-\theta_0$ then it is easily seen that for every real ξ and for every z in this region

$$\max.\frac{|\xi|}{|\xi - z|} = \frac{1}{\sin\theta_0}, \text{ so that } \frac{|\xi|}{|\xi - z|} < \frac{1}{\sin\theta_0} .$$

Hence

$$|F(\xi)| \leq \lambda\, |\xi|^{k} < \frac{\lambda}{\sin\theta_0}\, |\xi - z|^{k} = \lambda'(\theta_0)\, |\xi - z|^{k}.$$

Now

$$(2.5) \qquad |f_1(z)| = \left|\frac{1}{2\pi i} \int_{-\infty}^{\infty} \frac{f(\xi)}{(\xi - z)^2}\, d\xi\right|$$

$$= \frac{1}{2\pi} \left| \int_{-\infty}^{\infty} \frac{F(\xi)}{(\xi - z)^2}\, d\xi \right| < \frac{\lambda'(\theta_0)}{2\pi} \int_{-\infty}^{\infty} \frac{d\xi}{|\xi - z|^{2-k}}.$$

Again

$$(2.6) \qquad \int_{-\infty}^{\infty} \frac{d\xi}{|\xi - z|^{2-k}} = \left(\int_{-\infty}^{x} + \int_{x}^{\infty} \frac{d\xi}{|\xi - z|^{2-k}}\right) = \frac{2}{y^{1-k}} \int_{0}^{\pi/2} \cos^{-k}\phi\, d\phi$$

$$= \frac{C}{y^{1-k}} < \frac{C}{(r \sin\theta_0)^{1-k}}.$$

Now it follows from (2.5) and (2.6) that

$$|f_1(re^{i\theta})| < \frac{\lambda'(\theta_0)\ C}{2\pi\, (\sin\theta_0)^{1-k}\, r^{1-k}} = A(\theta_0)\left(\frac{1}{r^{1-k}}\right),$$

where $A(\theta_0)$ depends only on θ_0 and k. This proves (2.3).

For proving the inequality (2.4) first consider the function G(z) as defined in (1.10), since $0 < 1-k < 1$.

Thus

$$G(z) = \frac{1}{\sqrt{2\pi}} \int_{L} f_1(u) e^{-iuz}\, du,$$

where the ray L makes an angle ϕ with the real positive axis such that

---

1. For z lying in the upper half plane including the negative real axis we have $(1/2)\theta_0 < \pi - \phi < \theta$, and for z lying in the lower half plane including negative real axis we have $(1/2)\theta_0 < \phi < 2\pi - \theta$. Thus in either case, $\pi < \theta + \phi < 2\pi$ and $\sin(\theta + \phi)$ is negative.

$(1/2)\theta_0 < \phi < \pi-(1/2)\theta_0$. The position of L will be adjusted in the upper half plane according to the position of z in the region $\theta_0 < \theta < 2\pi-\theta_0$ so that $\sin(\theta+\phi)$ is negative.[1] Let $\sin(\theta+\phi) = -\mu$ where $\mu > 0$. Putting $u = Re^{i\phi}$ we obtain

$$G(re^{i\theta}) = \frac{1}{\sqrt{2\pi}} \int_0^\infty f_1(Re^{i\phi})e^{-irRe^{i(\theta+\phi)}} e^{i\phi}dR.$$

Using (2.3) we get

$$|G(re^{i\theta})| < \frac{A(\theta_0)}{\sqrt{2\pi}} \int_0^\infty \frac{1}{R^{1-k}} e^{-\mu rR}dR.$$

Hence

$$(2.7) \qquad |G(re^{i\theta})| < \frac{A(\theta_0)}{\sqrt{2\pi}} \frac{\Gamma(k)}{\mu^k r^k} < \frac{A(\theta_0)}{\sqrt{2\pi}} \cdot \frac{\Gamma(k)}{\mu_0^k r^k},$$

where $0 < \mu_0 \le \sin\frac{\theta_0}{2}$.

Now, having μ_0 such that $0 < \mu_0 < 1$, a constant C can be so chosen that $c\sin\theta_0 < \mu_0$. Consequently it follows from (2.7) that

$$(2.8) \qquad |G(re^{i\theta})| < A'(\theta_0)(1/r^k) \qquad \text{for } \theta_0 < \theta < 2\pi-\theta_0.$$

Similarly it can be shown that

$$(2.9) \qquad |H(re^{i\theta})| < A'(\theta_0)(1/r^k) \qquad \text{for } -\pi+\theta_0 < \theta < \pi-\theta_0.$$

Hence

$$|g_1(re^{i\theta})| = |H(re^{i\theta})-G(re^{i\theta})| < 2A'(\theta_0)(1/r^k), \quad \theta_0 < \theta < \pi-\theta_0$$

and

$$|g_2(re^{i\theta})| = |H(re^{i\theta})-G(re^{i\theta})| < 2A'(\theta_0)(1/r^k), \quad -\pi+\theta_0 < \theta < -\theta_0.$$

This proves (2.4) and Lemma 1 is established.

Corollary 1. If (f_1,f_2) be the function couple and (g_1,g_2) be the couple of CF-transform of a function $f \in L_p$ $(1 < p < \infty)$, then, since $f \in C'(1/q), 1/p+1/q=1$

$$(2.10)\qquad |f_1(re^{i\theta})| < A(\theta_o)\left(\frac{1}{r^{1/p}}\right) \qquad \text{for } \theta_o < \theta < \pi-\theta_o,$$

$$|f_2(re^{i\theta})| < A(\theta_o)\left(\frac{1}{r^{1/p}}\right) \qquad \text{for } -\pi+\theta_o < \theta < -\theta_o,$$

and

$$(2.11)\qquad |g_1(re^{i\theta})| < A_1(\theta_o)\left(\frac{1}{r^{1/q}}\right) \qquad \text{for } \theta_o < \theta < \pi-\theta_o,$$

$$|g_2(re^{i\theta})| < A_1(\theta_o)\left(\frac{1}{r^{1/q}}\right) \qquad \text{for } -\pi+\theta_o < \theta < -\theta_o,$$

for every θ_o such that $\theta_o < \theta < \pi/2$.

REMARK 1. It can be seen directly that for $f \in L_1(-\infty,\infty)$, the couple (f_1,f_2) satisfies the inequality (2.10) with p=1 and the couple (g_1,g_2) satisfies the inequality (2.11) with $q=\infty$, and $A_1(\theta_o)$ replaced by an absolute constant $||f||$ That is $g_1(z)$ and $g_2(z)$ are bounded in the upper half plane and the lower half plane respectively.

Note. The symbols $A(\theta_o)$, $A_1(\theta_o)$, $A'(\theta_o)$ and so on denote constants which depend only on θ_o and are not necessarily the same at every occurrence. The context will make it clear.

REMARK 2. If the Cf-transform couple (g_1,g_2) of a function $f \in L_p$ $(1<p<\infty)$ are defined as in (1.6) and (1.7), then by applying Holder's inequality to the integrals in (1.6) and (1.7) the inequalities (2.11) can be directly proved.

LEMMA 2. Let (f_1,f_2) be a couple of analytic functions holomorphic for Im(z) > 0 and Im(z) < 0 respectively and satisfying the condition

$$|f_1(re^{i\theta})| < A(\theta_o)\left(r^{\alpha}+\frac{1}{r^{\beta}}\right) \qquad \text{for } \theta_o < \theta < \pi-\theta_o,$$

(2.12)

$$|f_2(re^{i\theta})| < A(\theta_o)\left(r^{\alpha}+\frac{1}{r^{\beta}}\right) \qquad \text{for } \alpha \geq 0,\ \ 0 \leq \beta < 1,$$

for every θ_o such that $0 < \theta_o < \pi/2$, where $\alpha \geq 0$, $0 \leq \beta < 1$.

Then (g_1, g_2), the couple of CF-Transform of (f_1, f_2) satisfies the condition

(2.13)
$$|g_1(re^{i\theta})| < A_1(\theta_o)(r^{\beta-1} + \frac{1}{r^{\alpha+1}}) \quad \text{for } \theta_o < \theta < \pi-\theta_o,$$
$$|g_2(re^{i\theta})| < A_1(\theta_o)(r^{\beta-1} + \frac{1}{r^{\alpha+1}}) \quad \text{for } -\pi+\theta_o < \theta < -\theta_o,$$

for every θ_o such that $0 < \theta_o < \pi/2$.

PROOF. Let $G(z)$ be defined as in (1.10). Then as in the proof of Lemma 1, in the region of definition of G,

$$|G(re^{i\theta})| < \frac{A(\theta_o)}{\sqrt{2\pi}} \int_0^\infty (R^\alpha + \frac{1}{R^\beta})e^{-\mu r R} dR$$

$$= \frac{A(\theta_o)}{\sqrt{2\pi}} [\frac{\Gamma(\alpha+1)}{(\mu r)^{\alpha+1}} + \frac{\Gamma(-\beta+1)}{(\mu r)^{-\beta+1}}]$$

$$\leq \frac{A(\theta_o)}{\sqrt{2\pi}} \frac{\Gamma(\alpha+1)}{(\mu)^{\alpha+1}} [r^{\beta-1} + \frac{1}{r^{\alpha+1}}].$$

Hence it follows, as in (2.8), that

$$|G(re^{i\theta})| < A'(\theta_o)(r^{\beta-1} + \frac{1}{r^{\alpha+1}}) \quad \text{for } \theta_o < \theta < 2\pi-\theta_o. \tag{2.14}$$

In the same way such an inequality is proved for $H(re^{i\theta})$ which, when combined with (2.14) gives (2.13).

REMARK 3. It is evident from the proof of Lemma 2 that the conclusion (2.13) is still valid even if the numbers α and β assume negative values under the following restrictions :

(i) $-1 < \alpha < 0$, $0 \leq \beta < 1$, where $|\alpha| < \beta$,

or

(ii) $\alpha \geq 0$, $\beta < 0$, where $|\beta| < \alpha$.

It may be observed that the assumptions $|\alpha| < \beta$ in (i) and $|\beta| < \alpha$ in (ii) do not lead to any loss of generality.

LEMMA 3. Let (f_1,f_2) be a couple satisfying

$$|(f_1(re^{i\theta})| < A(\theta_o)(r^{\alpha} + \frac{1}{r^{\beta}}) \quad \text{for } \theta_o < \theta < \pi-\theta_o,$$

(2.15)

$$|(f_2(re^{i\theta})| < A(\theta_o)(r^{\alpha} + \frac{1}{r^{\beta}}) \quad \text{for } -\pi+\theta_o < \theta < \theta_o,$$

for every θ_o such that $0 < \theta_o < \pi/2$, where $\alpha > -1$, and $\beta \geq 0$, assuming furthe without loss of generality that $|\alpha| < \beta$ when α is negative and $0 < \beta < 1$. Then (g_1,g_2), the couple of CF transform of (f_1,f_2) satisfies the following inequalities :

$$|(g_1(re^{i\theta})| < A(\theta_o)(r^{\beta-1+\epsilon} + \frac{1}{r^{\alpha+1}}) \quad \text{for } \theta_o < \theta < \pi-\theta_o,$$

(2.16)

$$|(g_2(re^{i\theta})| < A(\theta_o)(r^{\beta-1+\epsilon} + \frac{1}{r^{\alpha+1}}) \quad \text{for } -\pi+\theta_o < \theta < -\theta_o,$$

for every θ_o such that $0 < \theta_o < \pi/2$, where $\epsilon > 0$.

PROOF. For $0 \leq \beta < 1$ the above result has already been proved in Lemma 2 with the only difference that the number ϵ does not figure in the conclusion (2.13). But this does not matter at all, since by multiplying $A_1(\theta_o)$ by a suitable constant the inequality (2.13) can always be put in the form of (2.16). Therefore we proceed to prove (2.16) for $\beta \geq 1$ assuming first that β is nonintegral. Choose a positive integer m such that $0 < \beta-m < 1$, i.e. $m = [\beta]$.

Now in this case G(z) will be determined from its derivative of mth order $G^{(m)}(z)$ which is given by (1.12). Using the same notations and taking the same ray L as was done in lemma 1 we have

(2.17) $$G^{(m)}(z) = \frac{(-i)^m}{\sqrt{2\pi}} \int_L u^m f_1(u) e^{-iuz} du = \frac{(-i)^m}{\sqrt{2\pi}} \left(\int_{L'} + \int_{L''}\right)$$

$$= G_1^{(m)}(z) + G_2^{(m)}(z),$$

where $L = L' + L''$, L' being that part of L which lies between the points 0 and $e^{i\phi}$.

Considering $G_1^{(m)}(z)$ first, we have

$$|G_1^{(m)}(z)| \le \frac{A(\theta_o)}{\sqrt{2\pi}} \int_0^1 R^m \left(R^\alpha + \frac{1}{R^\beta}\right) e^{R_I \sin(\theta+\phi)} dR$$

$$= \frac{A(\theta_o)}{\sqrt{2\pi}} \int_0^1 \left(R^{m+\alpha} + \frac{1}{R^{\beta-m}}\right) e^{-\mu TR} \, dR.$$

This shows that $G_1^{(m)}(z)$ is uniformly bounded in the region $\theta_o < \theta < 2\pi - \theta_o$. We consider $G(z)$ for z lying in this region. It will now be shown that in this region i.e. $\theta_o < \theta < 2\pi - \theta_o$

(2.18) $$|G_1^{(m)}(re^{i\theta})| = o(1/r^\delta) \text{ uniformly when } r \to \infty,$$

where $0 < \delta < m+1-\beta$.

Since $\beta - m < 1$, we also have $0 < \delta < 1$.

Now

(2.19) $$\int_0^1 \left(R^{m+\alpha} + \frac{1}{R^{\beta-m}}\right) r^\delta e^{-\mu rR} dR \le \int_0^1 \left(R^{m+\alpha} + \frac{1}{R^{\beta-m}}\right) \frac{\delta^\delta e^{-\delta}}{(\mu R)^\delta} \, dR,$$

Since the maximum value of $r^\delta e^{-\mu rR}$ for $0 < r < \infty$ occurs when $r = \delta/\mu R$. As the second integral in (2.19) is finite and does not depend upon r, it follows from the theorem of Dominated Convergence that the integral on the left of (2.19) tends to zero as $r \to \infty$. This proves (2.18).

The next result which must be proved is

(2.20) $$|G_1^{(m)}(re^{i\theta})| < M/r^\delta, \qquad M = M(\theta_o, \alpha, \beta).$$

From (2.18) it follows easily that there is a constant r_o such that

(2.21) $\quad |G_1^{(m)}(re^{i\theta})| < B/r^{\delta}, \quad$ for all $r \geq r_o$, where $B=B(\theta_o,\alpha,\beta)$.

Since $G_1^{(m)}(z)$ is uniformly bounded in the region under consideration, it follows that there is a constant $B_1 = B_1(\theta_o,\alpha,\beta)$ such that

(2.22) $\quad |G_1^{(m)}(re^{i\theta})| < B_1/r^{\delta}, \quad$ for $0 < r \leq 1$.

We also have

(2.23) $\quad |G_1^{(m)}(re^{i\theta})| < Cr_o^{\delta}/r^{\delta}, \quad$ for $1 \leq r \leq r_o$, $C = C(\theta_o,\alpha,\beta)$.

Taking into account the inequalities (2.21), (2.22) and (2.23) we get (2.20).

To evaluate G(z) we see that

$$G(z) = \frac{1}{(m-1)!}\int_0^z (z-u)^{m-1}G_1^{(m)}(u)du + \frac{1}{2\pi}\int_{L''} f_1(u)e^{-iuz}du + P(z),$$

where P(z) is a polynomial in z of degree (m-1) at most.

While taking majorant of G(z), P(z) can be omitted, since the majorant of P(z) is not greater that the majorant of other terms.

Hence

$$|G(re^{i\theta})| \leq \frac{M}{(m-1)!}\int_0^r r^{m-1}\frac{1}{r^{\delta}}\,dR + \frac{A(\theta_o)}{\sqrt{2\pi}}\int_1^{\infty}(R^{\alpha} + \frac{1}{R^{\beta}})e^{-\mu rR}dR$$

i.e.

$$|G(re^{i\theta})| \leq \frac{M}{(m-1)!}\,\frac{1}{(1-\delta)}\,r^{m-\delta} + \frac{A(\theta_o)}{\sqrt{2\pi}}\left[\frac{\Gamma(\alpha+1)}{(\mu r)^{\alpha+1}} + \frac{1}{\beta+1}\right]$$

$$< M_1(\theta_o)[r^{m-\delta} + \frac{1}{r^{\alpha+1}}]$$

$$= A_1(\theta_o)(r^{\beta-1+\epsilon} + \frac{1}{r^{\alpha+1}}), \text{ where } \epsilon > 0.$$

If β is an integer, after multiplying A (θ_o) by a suitable constant the first inequality in (2.15) can be replaced by

$$|f_1(re^{i\theta})| < A'(\theta_o)(r^{\alpha} + \frac{1}{r^{\beta+\epsilon'}})$$

where ϵ' is an arbitrarily chosen small positive number.

Treating the case of H(z) in the same way we get

$$|G(re^{i\theta})| < A_1(\theta_o)(r^{\beta-1+\epsilon} + \frac{1}{r^{\alpha+1}}) \quad \text{for } \theta_o < \theta < 2\pi-\theta_o,$$

(2.24)

$$|H(re^{i\theta})| < A_1(\theta_o)(r^{\beta-1+\epsilon} + \frac{1}{r^{\alpha+1}}) \quad \text{for } -\pi+\theta_o < \theta < \pi-\theta_o,$$

for every θ_o such that $0 < \theta_o < \pi/2$.

(2.16) follows from (2.24). The lemma is thus proved.

REMARK 4. Taking into account the inequalities (2.13) for the case when $\alpha > 0$, $\beta < 0$ and $|\beta| < \alpha$, it can be seen that the inequalities (2.16) are valid in this case also.

LEMMA 4. If $f \in C(k)$, $0 \le k < \infty$, then the function couple (f_1, f_2) of f satifies the inequalities

$$|f_1(re^{i\theta})| < A(\theta_o)(r^{k-1+\epsilon} + \frac{1}{r}) \quad \text{for } \theta_o < \theta < \pi-\theta_o,$$

(2.25)

$$|f_2(re^{i\theta})| < A(\theta_o)(r^{k-1+\epsilon} + \frac{1}{r}) \quad \text{for } -\pi+\theta_o < \theta < -\theta_o,$$

for every θ_o such that $0 < \theta_o < \pi/2$, where $0 \le \epsilon < 1$.

PROOF. Suppose first that k is non-integral and let p be a positive integer such that $0 < p-k+1 = \alpha < 1$. It is sufficient to prove the inequality (2.25) for f_1 only, since the proof for the inequality involving f_2 will be similar.

Let

$$F(\xi) = \int_0^{\xi} f(t)dt, \qquad z = x+iy = re^{i\theta}, \quad y > 0.$$

It is known [1, p.43] that the couple (f_1, f_2) is obtained from an analytic function $F_1(z)$, holomorphic in the finite z-plane cut along the real axis and that $F_1^{(p)}$, the pth order derivative of F_1, is given by

$$F_1^{(p)}(z) = \frac{p!}{2\pi i} \int_{-\infty}^{\infty} \frac{f(\xi)}{(\xi - z)^{p+1}} d\xi .$$

Now

$$F_1^{(p)}(z) = \frac{p!}{2\pi i} \int_{-\infty}^{\infty} \frac{f(\xi)}{(\xi - z)^{p+1}} d\xi = \frac{(p+1)!}{2\pi i} \int_{-\infty}^{\infty} \frac{F(\xi)}{(\xi - z)^{p+2}} d\xi$$

$$= \frac{(p+1)!}{2\pi i} \left[\left(\int_{-\infty}^{-A} + \int_{-A}^{A} + \int_{A}^{\infty} \frac{F(\xi)}{(\xi - z)^{p+2}} d\xi \right) \right]$$

$$= \frac{(p+1)!}{2\pi i} \int_{-A}^{A} \frac{F(\xi)}{(\xi - z)^{p+2}} d\xi + h_A(z),$$

where A is any positive constant.

Choosing A such that $\int_0^{\xi} |f(t)|dt < C|\xi|^k$ for $|\xi| \geq A$,

we have

$$\left| \int_A^{\infty} \frac{F(\xi)}{(\xi - z)^{p+2}} d\xi \right| \leq \int_A^{\infty} \frac{C|\xi|^k}{|\xi - z|^{p+2}} d\xi < \frac{C}{(\sin\theta_o)^k} \int_{-\infty}^{\infty} \frac{d\xi}{|\xi - z|^{p+2-k}}$$

$$= \frac{C}{(\sin\theta_o)^k} \int_{-\infty}^{\infty} \frac{d\xi}{|\xi - z|^{1+\alpha}}, \quad \text{as in the proof of Lemma 1,}$$

$$= \frac{C'}{(\sin\theta_o)^k} \frac{1}{y^{\alpha}} < C(\theta_o)(1/r^{\alpha}).$$

It can be shown similarly that

$$\left| \int_{-\infty}^{A} \frac{F(\xi)}{(\xi - z)^{p+2}} d\xi \right| < C(\theta_o)(1/r^{\alpha}).$$

Hence $h_A(z)$ is integrable on a line passing throug the origin and we have

$$(2.26) \quad f_1(z) = \frac{1}{2\pi i} \int_{-A}^{A} \frac{F(\xi)}{(\xi - z)^2} d\xi + \frac{1}{(p-1)!} \int_0^z (z-\zeta)^{p-1} h_A(\zeta) d\zeta + P_A(z),$$

where $P_A(z)$ is a polynomial in z of degree (p-1) at most and line of integration in the second integral is the straight line segment between the points 0 and z.

Now

$$\left|\frac{1}{(p-1)!}\int_0^z (z-\zeta)^{p-1}h_A(\zeta)d\zeta\right| \le \frac{C(\theta_o)}{(p-1)!}\int_0^r r^{p-1}\frac{1}{R^\alpha}dR = C'(\theta_o)(r^{p-\alpha})$$

$$(2.27)\qquad = C'(\theta_o)(r^{k-1}).$$

If M_A is the maximum of $|F(\xi)|$ in the closed interval [-A,A], then

$$\left|\frac{1}{2\pi i}\int_{-A}^{A}\frac{F(\xi)}{(\xi-z)^2}d\xi\right| \le \frac{M_A}{2\pi}\int_{-\infty}^{\infty}\frac{d\xi}{|\xi-z|^2} = \frac{M_A}{2\pi}\frac{C}{y}$$

$$(2.28)\qquad < M'(\theta_o)(1/r)$$

If A_p denotes the maximum of the absolute values of the coefficient in $P_A(z)$, then

$$(2.29)\qquad |P_A(z)| < p.A_p(1+r^{p-1}).$$

Taking into account the inequalities (2.27), (2.28), (2.29) and the relation (2.26) we get

$$(2.30)\qquad |f_1(re^{i\theta})| < A(\theta_o)(r^{k-1}+\frac{1}{r}),\quad \text{for } \theta_o < \theta < \pi-\theta_o.$$

It is thus seen that for k non-integral the inequalities (2.25) are satisfied with $\epsilon = 0$. If k is a positive integer, we could replace it by $k+\varepsilon$, where ϵ is an arbitrarily small non-negative number satisfying $0 \le \epsilon < 1$, since a function $f \in C(k)$ also belongs to $C(k+\epsilon)$. Consequently in this case also the couple (f_1,f_2) satisfies the relation (2.25).
This proves Lemma 4.

The following corollary, which will be needed later may be mentioned here :

COROLLARY 2. If (f_1,f_2) be the function couple corresponding to $f \in C(k)$, $0 \le k < 1$, then

$$|f_1(re^{i\theta})| < A(\theta_o)(\frac{1}{r^{1-k}}+\frac{1}{r})\qquad \text{for } \theta_o < \theta < \pi-\theta_o,$$

(2.31)

$$|f_2(re^{i\theta})| < A(\theta_o)(\frac{1}{r^{1-k}}+\frac{1}{r})\qquad \text{for } -\pi+\theta_o < \theta < -\theta_o,$$

for every θ_o such that $0 < \theta_o < \pi/2$.

This corollary is proved quite simply following the method of Lemma 4, since in this case the couple (f_1, f_2) are derived directly without involving their derivatives.

LEMMA 5. If (f_1, f_2) be the function couple of a function $f \in C(k)$, $0 \le k < \alpha$ then

(2.32) $$\lim_{y \to +0} y f_1(x+iy) = 0 \quad \text{and} \lim_{y \to +0} y f_2(x-iy) = 0$$

for every real x.

PROOF. It is sufficient to prove the lemma for f_1 only, since the proof for f_2 will be similar. Let $p = [k]$. Then $f_1^{(p)}(z)$, the pth order derivative of $f_1(z)$, is given by

$$f_1^{(p)}(x+iy) = \frac{p!}{2\pi i} \int_{-\infty}^{\infty} \frac{f(\xi)}{(\xi - x - iy)^{p+1}} d\xi$$

$$= \frac{p!}{2\pi i} \left[\left(\int_{-\infty}^{x-\delta} + \int_{x-\delta}^{x+\delta} + \int_{x+\delta}^{\infty} \right) \frac{f(\xi)}{(\xi - x - iy)^{p+1}} d\xi \right], \quad \delta > 0, \ x \text{ fixed},$$

$$= h_\delta(z) + \frac{p!}{2\pi i} \int_{x-\delta}^{x+\delta} \frac{f(\xi)}{(\xi - z)^{p+1}} d\xi ,$$

where $z = x+iy$ for a fixed x, i.e. z lies on the ordinate through x

Hence

(2.33) $$f_1(x+iy) = \frac{1}{(p-1)!} \int_x^z (z-\zeta) h_\delta(\zeta) d\zeta + \frac{1}{2\pi i} \int_{x-\delta}^{x+\delta} \frac{f(\xi)}{\xi - z} d\xi + P_\delta(z)$$

where $P_\delta(z)$ is a polynomial of degree (p-1) at most and the line of integration in the first integral is the straight line segment joining the point z to x.

For a fixed δ, $h_\delta(\zeta)$ is a continuous function on the line between z and x, hence the first integral in (2.33) defines a continuous function $Q(z,\delta)$. Also

$$\left| \int_{x-\delta}^{x+\delta} \frac{f(\xi)}{(\xi-z)}\, d\xi \right| \leq \frac{1}{y} \int_{x-\delta}^{x+\delta} |f(\xi)|\, d\xi .$$

Consequently it follows from (2.33) that

$$(2.34)\quad y|f_1(x+iy)| \leq y|Q(z,\delta)| + \frac{1}{2\pi} \int_{x-\delta}^{x+\delta} |f(\xi)|\, d\xi + y|P_\delta(z)| .$$

Since $f(\xi)$ is integrable in $(x-\delta, x+\delta)$ and Q and P_δ are continuous in z for a fixed $\delta > 0$, the expression on the right of (2.34) can be made arbitrarily small by first choosing δ and then making y tend to zero. This proves Lemma 5.

REMARK 5. Let $f \in L_p(-\infty,\infty)$, $1 \leq p < \infty$, and let (f_1, f_2) be the function couple of f. Then for $y > 0$

$$\lim_{y \to +0} y^{1/p} f_1(x+iy) = 0 \quad \text{and} \quad y^{1/p} f_2(x-iy) = 0.$$

The proof of this result is quite simple firstly because f_1 is directly defined without involving its derivative and secondly because for $p > 1$ an application of Holder´s inequality to the integral $\int_{x-\delta}^{x+\delta} \frac{f(\xi)}{(\xi-x-iy)}\, d\xi$ gives the result. When $p = 1$, it is still simpler since it is not required to use Holder´s inequality.

LEMMA 6. Let $g(z)$ be an analytic function, which is holomorphic in the finite z-plane cut along the closed finite interval $[-h', h]$, $h', h \geq 0$, of the real axis. For $z = re^{i\theta} = x+iy$, let

$$(2.35)\quad \lim_{r \to \infty} |g(re^{i\theta})| = 0 \quad \text{uniformly in the angles} \quad \begin{cases} \theta_0 < \theta < \pi-\theta_0, \\ -\pi+\theta_0 < \theta < -\theta_0, \end{cases}$$

for every θ_0 such that $0 < \theta_0 < \pi/2$,

$$(2.36)\quad |g(re^{i\theta})| \leq \frac{M}{|\sin\theta|^p} \left(1 + \frac{1}{r^\beta}\right) \quad \text{in the angles of (2.35),}$$

where $p \geq \beta \geq 0$, M being a constant.

Then g(z) is holomorphic at $z=\infty$ and it assumes the value 0 there.

PROOF. Without any loss of generality we may suppose that p is a positive integer in the inequality (2.36). At the origin we take an angle of measure 2α, $0 < \alpha < \pi/2$, in the right half plane, with the positive real axis as its bisector.

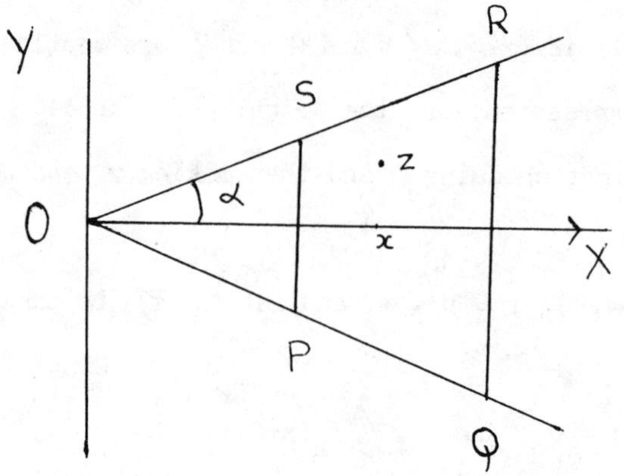

In this angle and lying in the region of holomorphy of g choose a point $z = x+iy$, with $x > h$. Take two ordinates SP and RQ terminated by the arms of the angle 2α, the abscissa of the first being greater than h and less than x and that of the second being equal to x+1. Denote the trapezium SPQR by Δ and its closure by $\bar{\Delta}$. Clearly g is holomorphic in $\bar{\Delta}$.

For a fixed z define a function $A(\zeta)$ by

(2.37) $$A(\zeta) = g(\zeta)(\zeta-x-1)^p,$$

where we take $\zeta = \xi+i\eta$ and $|\zeta| = \rho$.

Then the function $A(\zeta)$ is holomorphic in $\bar{\Delta}$.

If ζ is situated on the straight line RQ, then for sufficiently large values of x and consequently of ρ we have

$$|A(\zeta)| \le |g(\zeta)|\,|\eta|^p < M(\rho^\beta+1)\rho^{p-\beta} < M'\rho^p = \frac{M'(x+1)^p}{(\cos\alpha)^p},$$

where M′ is independent of x.

Hence for ζ lying on RQ

(2.38) $|A(\zeta)| < Kx^p$ for sufficiently large x,

where K is independent of x.

Since $g(\zeta)$ tends to zero as $|\zeta| \to \infty$ on the straight lines OR and OQ, and since the segment PS lies in the region of holomorphy of $g(\zeta)$, it follows that $|g(\zeta)|$ will be bounded when ζ lies on the boundary lines QP, PS and SR. Hence for ζ lying on these sides of Δ we get

(2.39) $|A(\zeta)| \le M_1(|\zeta| + x+1)^p < Cx^p$

for sufficiently large x, where C is independent of x.

If ζ lies on the boundary of Δ, it is clear from (2.38) and (2.39) that

(2.40) $|A(\zeta)| < C_1x^p \le C_1|z|^p$ for sufficiently large $|z|$, C_1 being independent of z.

Since $A(\zeta)$ satisfies the inequality (2.40) for any interior point of $\bar{\Delta}$, setting $\zeta = z$ we get

$$|A(z)| < C_1|z|^p \quad \text{for sufficiently large } z.$$

Hence for all sufficiently large values of z lying in the angle ROQ we have

(2.41) $$|g(z)| = \frac{|A(z)|}{|(z-x-1)|^p} < C_1|z|^p.$$

Since $\lim|g(z)| = 0$, as $|z| \to \infty$ on the lines OR and OQ, it follows from Phragmen-Lindelof theorem taking into account the condition (2.41) that

$$\lim_{|z| \to \infty} g(z) = 0 \quad \text{uniformly in the angle ROQ.}$$

It can similarly be proved that $\lim_{|z| \to \infty} g(z) = 0$, uniformly in a similar angle whose bisector is the negative real axis.

Hence, $\lim_{|z| \to \infty} g(z) = 0$ uniformly in the whole plane and consequently g(z) is holomorphic at $z=\infty$ and it takes the value zero there.

This proves Lemma 6.

REMARK 6. If in condition (2.35) we only assume that $|(g(z)| < A$, a constant, uniformly in the angles indicated there, then the conclusion will be that the function g is bounded at infinity and is holomorphic there.

A little modification of the proof of Lemma 6 will give the following result :

LEMMA 6a. Instead of satisfying the conditions (2.35) and (2.36) of Lemma 6, let the function g satisfy the following condition (2.35)a

(2.35)a As r approaches ∞, $|g(re^{i\theta})| < \dfrac{M}{|\sin\theta|^p}(r^\beta+1)$

in the angles indicated in (2.35). Then the singularity of the function g at the point at infinity is at most a pole of order $[\beta]$.

In fact, we consider the function $G(z) = \dfrac{g(z)}{(z)^{[\beta]}}$ which satisfies the condition of Lemma 6.

Using the transformation $w = 1/z$, we can adopt the conditions and conclusions of Lemmas 6 and 6a to get the following result :

LEMMA 6b. If g is holomorphic everywhere in the complex plane except perhaps at the points z=0 and z=∞, and if g satisfies the condition

(2.35)b $|g(re^{i\theta})| < \dfrac{M}{|\sin\theta|^p}\left(r^\alpha+\dfrac{1}{r^\beta}\right)$ in the angles of (2.35).

Then the function g has a pole of order $[\alpha]$ at most at the point $z = \infty$ and a pole of order $[\beta]$ at most at $z = 0$.

3. We are now in a position to prove some theorems, as indicated in the beginning of section 2. As before, (f_1,f_2) will denote the couple of analytic functions $f_1(z)$ and $f_2(z)$ which represent the function $f \in C(k)$, and (g_1,g_2) will represent the couple of analytic functions representing the

Fourier-Carleman transform of f or of (f_1,f_2).

THEOREM 1. *Let* $f \in C(k)$ *for* $0 \le k < 1$. *A necessary and sufficient condition for the function* f *to be the limit function of an analytic function* F(z), *which is holomorphic for* $\mathrm{Im}(z) > 0$ *and is represented by the Cauchy integral of* f, *is that*

(3.1) $$\lim_{y \to +0} \{g_1(x+iy)-g_2(x-iy)\} = 0$$ *uniformly in every closed finite interval* [a,b] *of the negative real axis*

i.e. $a \le x \le b < 0$.

(For the class $C'(k)$ $(0 \le k < 1)$, this theorem was announced in Singh [6]).

PROOF. (The condition is necessary). Let the function f be the limit function of F(z) which is holomorphic for $\mathrm{Im}(z) > 0$ and which is represented by the Cauchy integral of f. Then

$$F(z) = \frac{1}{2\pi i}\int_{-\infty}^{\infty} \frac{f(\xi)}{\xi - z}\, d\xi \qquad \text{for } \mathrm{Im}(z) > 0$$

$$= f_1(z) \qquad \text{for } \mathrm{Im}(z) > 0.$$

and $$\lim_{y \to +0} F(x+iy) = f(x) \text{ a.e.}$$

$$\lim_{y \to +0} f_1(x+iy) = f(x) \text{ a.e.}$$

But $$\lim_{y \to +0} \{f_1(x+iy)-f_2(x-iy)\} = f(x) \text{ a.e. as given in (1.5).}$$

Denoting by $f_o(z)$ the analytic function which is identically zero in the upper half plane we see that

$$\lim_{y \to +0} \{f_1(x+iy)-f_o(x-iy)\} = 0 \text{ a.e.}$$

Hence it follows from the uniqueness theorem of Carleman [1, page 47] that $f_1(z)-f_o(z)$ = a polynomial at most. But as $0 \le k < 1$, the polynomial reduces to zero and we have

$$f_2(z) \equiv 0, \qquad \text{for } \mathrm{Im}(z) < 0.$$

Now the couple (g_1,g_2) of the Carleman-Fourier transform of (f_1,f_2) are give by

$$(3.2)\qquad \begin{aligned} g_1(z) &= H(z)-G(z) &&\text{for } \mathrm{Im}(z) > 0,\\ g_2(z) &= H(z)-G(z) &&\text{for } \mathrm{Im}(z) < 0, \end{aligned}$$

where the function G(z) and H(z) are determined, because of the inequality (2.31), from their first derivatives $G^{(1)}(z)$ and $H^{(1)}(z)$ respectively. These derivatives are given by the integrals

$$G^{(1)}(z) = \frac{(-i)}{\sqrt{2\pi}} \int_L \xi f_1(\xi) e^{-i\xi z} d\xi ,$$

$$H^{(1)}(z) = \frac{(-i)}{\sqrt{2\pi}} \int_{L'} \xi f_2(\xi) e^{-i\xi z} d\xi ,$$

where the lines L and L´ are the same as in the integrals (1.10) and (1.11) respectively.

Since $f_2(z) \equiv 0$, it follows that $H^{(1)}(z) \equiv 0$.

Consequently $H(z) = C$, a constant.

Hence, taking into account (3.2) we have

$$\begin{aligned} g\ (z) &= C-G(z) &&\text{for } \mathrm{Im}(z) > 0,\\ g_2(z) &= C-G(z) &&\text{for } \mathrm{Im}(z) < 0. \end{aligned}$$

As explained in the paragraph following (1.13), G(z) is holomorphic on the negative real axis. Therefore

$\lim_{y \to +0} \{g_1(x+iy)-g_2(x-iy)\} = 0$ for all negative values of x, and since C-G(z) is holomorphic, this limit will be uniform in every closed finite interval [a,b] such that

$$a \le x \le b < 0.$$

Now we will take up the sufficiency part of the proof. Suppose that the condition (3.1) is satisfied. It follows from a theorem of Carleman [1, pages 38-39] that the functions $g_1(z)$ and $g_2(z)$ are analytic continuations of each other across the negative real axis. Hence we conclude from condition (3.2) that $H(z)-G(z)$ are holomorphic on and across the negative real axis. Since $G(z)$ is holomorphic on the negative real axis, it follows that the function $H(z)$ and its derivative $H^{(1)}(z)$ are holomorphic in the whole finite plane except perhaps at the origin.

Now we wish to show that

$$(3.3) \qquad |H^{(1)}(re^{i\theta})| < A'(\theta_o)\left(\frac{1}{r^{k+1}} + \frac{1}{r}\right) \qquad \text{for } -\pi+\theta_o < \theta < \pi-\theta_o$$

for every θ_o such that $0 < \theta_o < \pi/2$, where

$$A'(\theta_o) \text{ has the form } \frac{A'}{|\sin\theta o|^p}.$$

Since

$$H^{(1)}(z) = \frac{(-i)}{\sqrt{2\pi}} \int_{L'} \xi f_2(\xi) e^{-i\xi z} d\xi ,$$

where the ray L' is choosen according to the position of the point z as has been explained in Lemma 1. Setting $\xi = Re^{i\phi}$ and $\sin(\theta+\phi) = -\mu$, $\mu > 0$, we get for $\pi-\theta_o < \theta < \pi+\theta_o$,

$$|H^{(1)}(re^{i\theta})| < \frac{A(\theta_o)}{\sqrt{2\pi}} \int_0^\infty R(R^{k-1} + \frac{1}{R})e^{-\mu rR} dR$$

$$= \frac{A(\theta_o)}{\sqrt{2\pi}} \left[\frac{\Gamma(k+1)}{(\mu r)^{k+1}} + \frac{1}{\mu r}\right]$$

$$< A'(\theta_o)\left(\frac{1}{r^{k+1}} + \frac{1}{r}\right).$$

This proves (3.3). It is also clear from the above analysis and from Lemma 4 that $A'(\theta_o)$ has the form $\dfrac{A'}{|\sin\theta_o|^p}$. Taking into account the inequality (3.3) and applying Lemma 6b to $H^{(1)}(z)$ we see that $H^{(1)}(z)$ has at most a simple pole at the origin since $0 \leq k < 1$. again, using (3.3) and applying Lemma 6 we find that $H^{(1)}(z)$ is analytic at the point $z = \infty$ and takes the value zero there.

Hence

(3.4) $H^{(1)}(z) = B/z$, where B is a constant.

It follows from (3.4) that

(3.5) $zf_2(z) = C$, for every z in the lower half plane, where C is a constant.

Now making z tend to zero along the negative imaginary axis it follows from Lemma 5 that $\lim_{y \to +0} (-iy)f_2(-iy) = 0$.

Consequently C=0 in (3.5), and we conclude that $f_2(z) \equiv 0$.

Hence it follows that $\lim_{y \to +0} f_1(x+iy) = f(x)$ a.e., where for y > 0,

$$f_1(x+iy) = f_1(z) = \frac{1}{2\pi i} \int_{-\infty}^{\infty} \frac{f(\xi)}{\xi - z} d\xi .$$

This proves Theorem 1.

REMARK 7. Theorem 1 is a generalization of a theorem of Hille and Tamarkin [2]. The proof of the theorem for the case when $f \in C'(k)$, $0 \leq k < 1$ is simpler. We have also proved this theorem for the general case when $f \in C(k)$, $0 \leq k < \infty$.

4. The following theorem is a generalization of another theorem of Hille and Tamarkin [3].

THEOREM 2. Let $f \in C(k)$, $(0 \leq k < 1)$ and let (f_1, f_2) be the functional couple of f. Then there exists a couple (g_1, g_2) whose Carleman-Fourier transform is the couple (f_1, f_2). A necessary and sufficient condition for the function f(x) to be the limit function of an analytic function F(z), holomorphic for Im(z) > 0 and represented by the Cauchy integral of f, is that

(4.1) $\lim_{y \to +0} \{g_1(x+iy) - g_2(x-iy)\} = 0$ uniformly in every finite closed interval of the positive real axis.

(Theorem 2 was announced in Singh, U.N. [6]).

PROOF. The proof of this theorem is omitted since it is similar to the proof of Theorem 1 except that the following inversion formula for C-F transform is used in the proof :

(4.2) $$g \equiv TST(f).$$

5. The following theorem is a generalization of a theorem of Paley and Wiener, [4, pp.16-17].

THEOREM 3. *Let $\phi(x)$ be ≥ 0, but $\not\equiv 0$ on the real line and let $\phi \in L_p(-\infty,\infty)$ for $1 \leq p < \infty$. Then the condition*

(5.1) $$\int_{-\infty}^{\infty} \frac{|\log \phi(x)|}{1+x^2}\, dx < \infty$$

is necessary and sufficient for the existence of a function $f \in L_p(-\infty,\infty)$ such that

(i) $|f(x)| = \phi(x)$ *almost everywhere*,

and

(ii) $\lim_{y \to +0} \{g_1(x+iy)-g_2(x-iy)\} = 0$ *uniformly in every closed finite interval* $[a,b]$ *of the negative real axis*, i.e. $a \leq x \leq b < 0$.

(Paley and Wiener proved this theorem for the class L_2 and Hille and Tamarkin proved it for the class L_p, $1 \leq p < 2$).

PROOF. (The condition is necessary). Suppose that there is a function $f \in L_p$ which satisfies the conditions (i) and (ii) of the theorem. Then it follows from theorem 1 that there exists a function $F(z)$, holomorphic for $Im(z) > 0$ and such that it is represented by the Cauchy integral of f and $f(x)$ is its limit function. In fact, the function $F(z)$ is no other function than $f_1(z)$. Let us transform the upper plane $Im(z) > 0$ of the z-plane into the unit disk $|w| < 1$ of the w-plane by the transformation

$$w = (1+iz)/(1-iz).$$

Then the real axis of the z-plane will be transformed into the circumference of the unit circle in the w-plane. Under this transformation the point $\zeta = e^{i\psi}$ of the circumference of the unit circle in the w-plane will correspond to the point $x=\tan\psi/2$ of the real axis of the z-plane. Further suppose that the function $F(z)$ is transformed into the function $s(w)$ and that the limit function of $s(w)$ is the function $S(\zeta)$.

Then $f(x) = S(\zeta) = S(e^{i\psi})$.

Since $F(z)$ is represented by Cauchy integral of f, and as $f_2(z)\equiv 0$, we have for $y > 0$

$$F(x+iy) = f_1(x+iy)-f_2(x-iy) = \frac{1}{\pi}\int_{-\infty}^{\infty} \frac{f(\xi)y}{(\xi-x)^2+y^2}\, d\xi ,$$

that is to say that $F(z)$ is also represented by the Poisson integral of f. Hence it follows that the function $s(w)$ also will be represented by the Cauchy integral as also the Poisson integral of its limit function $S(\zeta)$. This will imply that $\log|S(e^{i\psi})|$ is integrable [2] in $(-\pi,\pi)$.

Consequently

$$\int_{-\infty}^{\infty} \frac{|\log\phi(x)|}{1+x^2}\, dx = \int_{-\infty}^{\infty} \frac{|\log|f(x)||}{1+x^2}\, dx = \frac{1}{2}\int_{-\pi}^{\pi} |\log|S(e^{i\psi})||\, d\psi < \infty.$$

We now take up the proof of the sufficiency part of the condition (5.1). Let the condition (5.1) be satisfied.

Then for $z = x+iy$, $y > 0$,

$$\lambda(x,y) = \frac{1}{\pi}\int_{-\infty}^{\infty} \frac{\log\phi(\xi)y}{(\xi-x)^2+y^2}\, d\xi$$

is a harmonic function in the upper half plane and

(5.2) $$\lim_{y\to+0} \lambda(x,y) = \log\phi(x) \quad \text{almost everwhere.}$$

Denote by $\mu(x,y)$ the function conjugate to $\lambda(x,y)$ and set

$$h(z) = e^{\lambda(x,y)+i\mu(x,y)}.$$

Then h(z) is holomorphic in the upper half plane and we have

$$(5.3) \qquad \lim_{y \to +0} |h(z)| = \phi(x) \quad \text{a.e.}$$

Now

$$|h(x+iy)| = e^{\lambda(x,y)} = \exp\{\frac{1}{\pi}\int_{-\infty}^{\infty} \frac{\log\phi(\xi)y}{(\xi-x)^2+y^2}d\xi\}.$$

But

$$\exp\{\frac{1}{\pi}\int_{-\infty}^{\infty} \frac{\log\phi(\xi)y}{(\xi-x)^2+y^2}d\xi\} = \exp\{\frac{\int_{-\infty}^{\infty} \frac{\log\phi(\xi)y}{(\xi-x)^2+y^2}d\xi}{\int_{-\infty}^{\infty} \frac{yd\xi}{(\xi-x)^2+y^2}}\}$$

$$\leq \frac{\int_{-\infty}^{\infty} \frac{\log\phi(\xi)y}{(\xi-x)^2+y^2}d\xi}{\int_{-\infty}^{\infty} \frac{yd\xi}{(\xi-x)^2+y^2}}$$

by Jensen's inequality [7], and

$$\frac{\int_{-\infty}^{\infty} \frac{\log\phi(\xi)y}{(\xi-x)^2+y^2}d\xi}{\int_{-\infty}^{\infty} \frac{yd\xi}{(\xi-x)^2+y^2}} = \frac{1}{\pi}\int_{-\infty}^{\infty} \frac{\phi(\xi)y}{(\xi-x)^2+y^2}d\xi .$$

Hence

$$(5.4) \qquad |h(x+iy)| \leq \frac{1}{\pi}\int_{-\infty}^{\infty} \frac{\phi(\xi)y}{(\xi-x)^2+y^2}d\xi$$

Consider the case $p > 1$ first and suppose that q is the conjugate index of p i.e. $1/p + 1/q = 1$. Then applying Holder's inequality to the r.h.s. of (5.4) we get

$$|h(x+iy)| \le \frac{1}{\pi}\left[\int_{-\infty}^{\infty} \frac{[\phi(\xi)]^p y}{(\xi-x)^2+y^2}\, d\xi\right]^{1/p} \left[\int_{-\infty}^{\infty} \frac{y}{(\xi-x)^2+y^2}\, d\xi\right]^{1/q}.$$

Therefore,

$$|h(x+iy)|^p \le \frac{1}{\pi^p}\int_{-\infty}^{\infty} \frac{[\phi(\xi)]^p y}{(\xi-x)^2+y^2}\, d\xi \,.\, \pi^{p/q}$$

$$= \frac{1}{\pi}\int_{-\infty}^{\infty} \frac{[\phi(\xi)]^p y}{(\xi-x)^2+y^2}\, d\xi\,.$$

Consequently

$$\int_{-\infty}^{\infty} |h(x+iy)|^p dx \le \frac{1}{\pi}\int_{-\infty}^{\infty} dx \int_{-\infty}^{\infty} \frac{[\phi(\xi)]^p y}{(\xi-x)^2+y^2}\, d\xi$$

$$= \frac{1}{\pi}\int_{-\infty}^{\infty} [\phi(\xi)]^p d\xi \int_{\infty}^{\infty} \frac{y\, dx}{(\xi-x)^2+y^2}$$

$$= \frac{1}{\pi}\int_{-\infty}^{\infty} [\phi(\xi)]^p d\xi = M_1, \text{ a constant.}$$

If p=1, we get from (5.4)

$$\int_{-\infty}^{\infty} |h(x+iy)|\, dx \le \frac{1}{\pi}\int_{-\infty}^{\infty} dx \int_{-\infty}^{\infty} \frac{\phi(\xi).y}{(\xi-x)^2+y^2}\, d\xi$$

$$= \frac{1}{\pi}\int_{-\infty}^{\infty} \phi(\xi) d\xi \int_{\infty}^{\infty} \frac{y\, dx}{(\xi-x)^2+y^2}$$

$$= \frac{1}{\pi}\int_{-\infty}^{\infty} \phi(\xi) \quad \xi = M_2, \text{ a constant.}$$

Hence for $1 \le p < \infty$, we have

$$\text{(5.5)} \qquad \int_{-\infty}^{\infty} |h(x+iy)|^p dx \le M,$$

which shows that $h(z) \in H_p$, and it follows from this fact that $\lim_{y \to +0} h(x+iy)$ exists for almost all values of x. Denoting this limit function by f(x) we observe, [4], that $f(x) \in L_p(-\infty,\infty)$ and that h(z) is represented by the Cauch integral of f.

If (g_1,g_2) is the couple of Fourier-Carleman transform of f, then it follows from theorem 1 that

$$\lim_{y \to +0} \{g_1(x+iy)-g_2(x-iy)\} = 0$$ uniformly in every closed finite interval

$a \leq x \leq b < 0$ of the negative real axis.

Also using the fact that f(x) is the limit function of h(z) and the relation (5.3) we get

$$\phi(x) = \lim_{y \to +0} |h(x+iy)| = |f(x)| \quad \text{a.e.}$$

This proves the theorem.

REFERENCES

1. Carleman, T., L'intégrale de Fourier et questions qui s'y rattachent, Uppsala (1944).

2. Hille, E. and Tamarkin, J.D., On a theorem of Paley and Wiener, Annals of Mathematics (2), 34(1933), 606-614.

3 Hille, E. and Tamarkin, J.D., A remark on Fourier transform and functions analytic in a half plane, Compositio math. (1934), 98-102.

4. Paley, R.E.A.C., and Wiener, N., Fourier transforms in the complex domain, A.M.S. Colloquium Publications, New York (1934).

5. Singh, U.N., Sur quelques théoremes de Hille et Tamarkin, Comptes Rendus, Paris 236(1953), 885-887.

6. Singh, U.N., Fonctions entières et transformée de Fourier généralisée, Comptes Rendus, Paris 237(1953), 14-16.

7. Zygmund, A., Trignometric Series, Second edition, Vols I and II, Cambridge University Press (1968).

M.P. University Grants Commission
E-2/84, Arera Colony
Bhopal-462016
India.

EXTENDING SEMINORMS IN LOCALLY PSEUDOCONVEX ALGEBRAS

W.ZELAZKO

DEDICATED TO THE MEMORY OF U.N. SINGH

A locally convex algebra A is a real or complex associative algebra, which is a locally convex Hausdorff topological vector space and the multiplication (x,y) -> xy is a jointly continuous bilinear operation from AxA to A. The topology of a locally convex algebra A can be given by means of a family $\Phi = (1.1_\alpha)_{\alpha\in\mathfrak{a}}$ of seminorms such that for each index α there is a $\beta\in\mathfrak{a}$ with

(1) $$|xy|_\alpha \leq |x|_\beta |y|_\beta$$

for all x,y in A. Without loss of generality it can be assumed that Φ has the following property: together with a finite set of indices $\{\alpha_1, \ldots, \alpha_n\}$ the (continuous) seminorm

(2) $$|x| = \max\{|x|_{\alpha_1}, \ldots, |x|_{\alpha_n}\}$$

is in Φ too. This property has a technical character and for the purpose of this paper we call it property (m). It was a long standing problem posed in [4] (see also [5,p 78]), whether in the case when A possesses a unit element e the seminorm in Φ can be chosen so that

(3) $$|e| = 1$$

for all α in $\mathfrak{a}$.The problem has been recently solved in the affirmative by Fernández and Müller [1] who proved that if A_0 is a subalgebra of a locally convex algebra A then any system Φ_0 of seminorms on A_0 satisfying (1), having the property (m) and giving the (relative) topology of A_0, can be extended to a system Φ of seminorms on A, giving its topology and satisfying (1). More precisely, Φ_0 equals to the family of restrictions to A of the member of Φ. If A has a unit element e, then taking as A_0 the scalar multiples of the unit λe and setting $|\lambda e|=|\lambda|$, what is a norm on A_0, we see that the system Φ obtained in the theorem of Fernandez and Muller satisfies both (1) and (3).

The purpose of this paper is to modify the proof of the above result, so that it can be applied to a more general class of locally pseudoconvex algebras. While the result cannot be literally extended to this class (see remark at the end of the paper) some its variation is still true and it is strong enough in order to obtain formulas analogous to (1) and (3).

Let X be a real or complex vector space. A non-negative function $x \to ||x||$ on X is said to be a p-homogeneous seminorm, or a seminorm with exponent p, $0<p\leq 1$, if

(i) $||x+y|| \leq ||x||+||y||$

for all $x,y\in X$, and

(ii) $||\lambda x||=|\lambda|^p\,||x||$

for all scalars λ and all elements x in X.

Note that if $||x||$ is a seminorm of exponent p, and q is a real number satisfying $0<q\leq 1$, then $||x||^q$ is a seminorm of exponent pq. The relation (i) follows here immediately from the inequality $(1+|\lambda|)^q\leq 1 + |\lambda|^q$, what implies immediately

$$(4) \qquad (\sum_{i=1}^{n} |\lambda_i|)^q \leq \sum_{i=1}^{n} |\lambda_i|^q$$

for any n-tuple of scalars $\lambda_1, \dots ,\lambda_n$.

A locally pseudoconvex space X (see [2, Chapter 3], or [3,p 4]) can be described as a topological vector Hausdroff space whose topology is given by the means of a family $\phi(x)=(1.1_\alpha)_{\alpha\in a}$ of $p(\alpha)$-homogeneous seminorms, $0<p(\alpha)\leq 1$. That means that a net (x_i) tends to zero in X whenever $||x_i||_\alpha \underset{i}{\to} 0$ for each fixed α in a.

Relations (i) and (ii) imply that the operations of addition and scalar multuiplication are continuous in this topology. A locally pseudoconvex algebra is a real or complex associative algebra which is a locally pseuduconvex space and the multiplication is a jointly continuous bilinear operation. Similarly as in the locally convex case it can be shown that the topology of a locally pseudoconvex algebra A can be given by means of a family $\phi(A)=(1.1_\alpha)_{\alpha\in a}$ of $p(\alpha)$-homogenuous seminorms, $0<p(\alpha)\leq 1$, such that for each $\alpha\in r$ there is a $\beta\in r$ with

$$(5)\qquad |xy|_{\alpha}^{p(\beta)} \leq |x|_{\beta}^{p(\alpha)}\,|y|_{\beta}^{p(\alpha)}$$

for all $x,y\in A$.

EXAMPLE. Let $a_i^{(n)}$, $n=1,2,\ldots,i=0,1,\ldots$ be an infinite matrix of positive real numbers, and let (q_i) be a sequence of numbers satisfying $0<q\leq 1$, such that

$$[a_{i+j}^{(n)}]^{q_{n+1}} \leq a_i^{(n+1)}\,a_j^{(n+1)}$$

for all involved i,j,n. Put $p_1=q_1, p_n=q_n.P_{n-1}$ for $n>1$ and define A to be the set of all (real or complex) sequences $x=(\xi_i)_o^{\infty}$ such that

$$(6)\qquad ||x||_n = \sum_{i=o}^{\infty} a_i^{(n)}\,|\xi_i|^{p_n} < \infty$$

for $n=1,2,\ldots$ One can easily verify that the formula (6) defines on A a sequence of P_n-homogeneous seminorms, so that A becomes a pseuduconvex space under the family $\phi=(||.||_n)_o^{\infty}$. Define on A the convolution multiplication: if $x=(\xi_n), y=(\nu_n)$, then $xy=z=(\zeta_n)$, where

$\zeta_k = \sum_{i=0}^{\infty} \xi_{k-i}\nu_i$ (writing $\xi_j, \nu_j = 0$ for $j<0$). Using (4) we obtain

$$||xy||_n^{q_{n+1}} = \Big(\sum_{k=o}^{\infty} a_k^{(n)}\,\Big|\sum_{i=o}^{\infty}\xi_{k-i}\nu_i\Big|^{p_n}\Big)^{q_{n+1}} \leq$$

$$\sum_{k=o}^{\infty}\sum_{i=o}^{\infty} [a_k^{(n)}]^{q_{n+1}}\,|\xi_{k-i}|^{p_n q_{n+1}}|\nu_i|^{p_n q_{n+1}} \leq$$

$$\leq \sum_{k=o}^{\infty}\sum_{i=o}^{\infty} a_{k-i}^{(n+1)}a_i^{(n+1)}|\xi_{k-i}|^{p_{n+1}}|\nu_i|^{p_{n+1}} = ||x||_{n+1}||y||_{n+1}.$$

Taking the r_n-th powers on the both sides we obtain (5) and so A is a locally pseudoconvex algebra. A concrete example can be obtained by taking $a_i^{(n)} = \exp(i2^{-n})$ and $q_i=1/2$. It is easy to see that the algebra of this example is not locally convex.

Similarly as in the locally convex case we say that the system $\phi=(|.|_{\alpha})_{\alpha\in a}$ of $p(\alpha)$-homogeneous seminorms has the property (m) if for each finite n-tuple of indices $\alpha_1, \ldots, \alpha_n$ the seminorm

$$(7)\qquad |x| = \max\{|x|^1, \ldots, |x|^n\}$$

is in ϕ. Here $s_i=s/p(\alpha_i)$ and $s=p(\alpha_1)...p(\alpha_n)$, so that s is the exponent of the seminorm given by (7). Since all seminorms of the form (7) are continuous, we can assume, without loss of generality that ϕ has the property (m). Otherwise we can add to ϕ all seminorms of the form (7). The new system will also satisfy relations (5) because the seminorm of the form (7) applied to a product xy can be estimated from above by the product $|x|_1|y|_1$, where $|.|_1$ is again a seminorm of the form (7) but with α_i replaced by β_i, where α_i is associated to β_i by the formula (5).

The property (m) implies that if 1.1 is a continuous p-homogeneous seminorm on A, then there is an index $\alpha\in a$ and a positive C such that

$$(8) \qquad |x|^{p(\alpha)} \leq C|x|_\alpha^p$$

for all x in A, and we need property (m) in order to make use of this formula.

In the case we have two systems of seminorms ϕ and ϕ_1 on A and each seminorm in ϕ is continuous with respect to the topology given by ϕ_1, we write $\phi \leq \phi_1$. We call ϕ and ϕ_1 to be equivalent if both $\phi \leq \phi_1$ and $\phi_1 \leq \phi$. It is clear that two systems ϕ and ϕ_1 are equivalent if and only if they define the same topology.

We can formulate now the main result of this paper.

THEOREM. Let A be a real or complex locally psedoconvex algebra, and let A_o be its subalgebra. Let $\phi_o=(||.||_\alpha)_{\alpha_o}$ be a system of $p_o(\alpha)$-homogeneous seminorms on A_o, which gives its (relative) topology, satisfies (5) and has the property (m). Then the topology of A can be given by means of a family $\phi_1=(|.|_\alpha)_{\alpha\in a_1}$ of $(p_1(\alpha)$-homogeneous seminorms satisfying condition (5), such that the restriction of each seminorm $|.|_\alpha \in \phi_1$ to A_o coincides with some power of a seminorm in ϕ_o, with exponent q_α satisfying $0 < q_\alpha \leq 1$.

PROOF. Let $\phi=(|.|_\alpha)_{\alpha\in a}$ be a system of $p(\alpha)$-homogeneous seminorms on A, giving its topology, satisfying condition (5) and possesing the property (m). Since the restrictions of seminorms in Φ to A_o give its topology and have also the property (m), then by (8) for each α in a_o there is a β in a and a positive constant C such that

(9) $$||a||_{\alpha}^{p_1(\beta)} \leq C|a|_{\beta}^{p_o(\alpha)}$$

for all a in A_o. Denote by r_1 the family of all quadruples (α,β,C,r), where $\alpha\in a_o$, $\beta\in a$, $c>0$, $o<r\leq 1$, and α together with β and C satisfy relation (8). For $\gamma=(\alpha,\beta,C,r)\in a_1$ and $x\in A$ define

(10) $$|x|_{\gamma}^{(1)} = \inf\{||a||_{\alpha}^{rp(\beta)} + C^r|x-a|_{\beta}^{rp_o(\alpha)} : a\in A_o\}.$$

It is easy to see that the formula (10) defines on A a seminorm with exponent $P_1(\gamma)=rp_o(\alpha)p(\beta)$. Define $\Phi_1=\{|.|_{\gamma}^{(1)}\}_{\gamma\in a_1}$. We shall prove that Φ_1 is the desired family.

Fix an index $\gamma = (\alpha,\beta,C,r)$ in a_1 and x in A_o. Using relation (9) and triangle inequality for the seminorm $||.||_{\alpha}^{rp(\beta)}$ we obtain

$$||a||_{\alpha}^{rp(\beta)}+C^r|x-a|_{\beta}^{rp_o(\alpha)} \geq ||a||_{\alpha}^{rp(\beta)} + ||x-a||_{\alpha}^{rp(\beta)} \geq ||x||_{\alpha}^{rp(\beta)}$$

for every a in A_o. Taking the infimum of the left-hand expression with respect to a in A_o we obtain

(11) $$|x|_{\gamma}^{(1)} \geq ||x||_{\alpha}^{rp(\beta)}$$

and this holds true for each x in A_o. Setting a=x in the right-hand expression in the formula (10) we obtain

(12) $$|x|_{\gamma}^{(1)} \leq ||x||_{\alpha}^{rp(\beta)}$$

for every x in A_o. Thus by (11) and (12) we have $|x|_{\gamma}^{(1)}=||x||_{\alpha}^{rp(\beta)}$ for x in A_o, and so the restrictions to A_o of seminorms in Φ_1 are powers of elements in Φ_o with exponents $rp(\beta)$ satisfying $0<rp(\beta)\leq 1$. It remains to be shown that both systems Φ and Φ_1 are equivalent and that Φ_1 satisfies relations (5).

Setting a=0 in the right-hand expression of the formula (10) we obtain immediately

$$|x|_\gamma^{(1)} \leq C^r |x|_\beta^{r p_o(\alpha)}$$

for all x in A and all indices γ in α_1. Thus $\phi_1 \leq \phi$. We fix now an index δ in α. By the property (m) of ϕ_o we can find an index α in α_o and a positive constant C_o so that

(13) $\qquad |a|_\delta^{p_o(\alpha)} \leq C_o ||a||^{p(\delta)}$

for all a in A_o. Using the property (m) of the system ϕ we can find a β in α so that relation (9) is satisfied together with α for some positive constant C, and moreover

(14) $\qquad |x|_\delta^{p(\beta)} \leq C_1 |x|_\beta^{p(\delta)}$

is satisfied for all x in A, where C_1 is a fixed positive number. Let $\gamma = (\alpha, \beta, C, p(\delta))$. By the formula (9) we have $\gamma \in \alpha_1$. Let $a \in A_o$ and $x \in A$. By (13) and (14) we have

$$||a||_\alpha^{p(\beta)p(\delta)} + C^{p(\delta)}|x-a|^{p_o(\alpha)p(\delta)} \geq$$

$$\geq C_o^{-p(\beta)}|a|_\delta^{p_o(\alpha)p(\beta)} + C^{p(\delta)}C_1^{-p_o(\alpha)}|x-a|_\delta^{p_o(\alpha)p(\beta)} \geq$$

$$\geq C_2 |x|_\delta^{p_o(\alpha)p(\beta)},$$

where $C_2 = \min\{C_o^{-p(\beta)}, C^{p(\delta)}C_1^{-p_o}\}$.

Taking the infimum of the left-hand expression with respect to a in A_o we obtain by (10)

$$|x|_\gamma^{(1)} \geq C_2 |x|_\delta^{p_o(\alpha)p(\beta)}$$

for all x in A. Since δ was chosen arbitrarily in α, we obtain $\phi \leq \phi_1$. Thus both systems ϕ and ϕ_1 are equivalent. We shall be done when we show that the system ϕ_1 satisfies relation (5). Fix a $\gamma = (\alpha, \beta, C, r)$ in α_1. Since ϕ satisfies (5) we can find a ν in α so that

(15) $\qquad |xy|_\beta^{p(\nu)} \leq |x|_\nu^{p(\beta)} |y|_\nu^{p(\beta)}$

for all x,y in A. Using (5), the property (m) of ϕ_o and the fact that the restrictions to A_o of seminorms in ϕ give a system equivalent to ϕ_o, we find a μ in α_o and a constant $C_1 > 0$ so that

(16) $$|a|_{\nu}^{p_o(\mu)} \le C_1 ||a||_{\mu}^{p(\nu)}$$

(17) $$||ab||_{\alpha}^{p_o(\mu)} \le ||a||_{\mu}^{p_o(\alpha)} ||b||_{\mu}^{p_o(\alpha)}$$

for all a,b in A_o. Moreover, by (9) and the property (m) of ϕ we find a positive C_2 and an index ρ in α such that

(18) $$||a||_{\mu}^{p_o(\rho)} \le C_2 |a|_{\rho}^{p_o(\mu)}$$

for all a in A_o, and

(19) $$|x|_{\nu}^{p(\rho)} \le |x|_{\rho}^{p(\nu)}$$

for all x in A.

We fix arbitrarily x,y in A and a,b in A_o. Using definition (10),
the formula

$$xy = ab + a(y-b) + (x-a)b + (x-a)(y-b)$$

and relations (4), (15),(16),(17) and (19) we can write

$$[|xy|_{\gamma}^{(1)}]^{p_o(\mu)p(\nu)p(\rho)}$$

$$\le (||ab||_{\alpha}^{r\,p(\beta)} + C^r |a(y-b)+(x-a)b+(x-a)(y-b)|_{\beta}^{rp_o(\alpha)})^{p_o(\mu)p(\nu)p(\rho)}$$

$$\le (||ab||_{\alpha}^{rp(\beta)p_o(\mu)p(\nu)p(\rho)} + C^{rp_o(\mu)p(\nu)p(\rho)}[|a(y-b)|_{\beta}^{rp_o(\alpha)p_o(\mu)p(\nu)p(\rho)}$$

$$+ |(x-a)b|_{\beta}^{rp_o(\alpha)p_o(\mu)p(\nu)p(\rho)} + |(x-a)(y-b)|^{rp_o(\alpha)p_o(\nu)p(\rho)}$$

$$\le ||a||_{\mu}^{rp(\beta)p_o(\alpha)p(\nu)p(\rho)} ||b||_{\mu}^{rp(\beta)p_o(\alpha)p(\nu)p(\rho)}$$

$$+ C^{rp_o(\mu)p(\nu)p(\rho)}[|a|_{\nu}^{rp_o(\alpha)p_o(\mu)p(\beta)p(\rho)} |y-b|_{\nu}^{rp_o(\alpha)p_o(\mu)p(\beta)p(\rho)} +$$

$$+ |x-a|_{\nu}^{rp_o(\alpha)p_o(\mu)p(\beta)p(\rho)} |b|_{\nu}^{rp_o(\alpha)p_o(\mu)p(\beta)p(\rho)} +$$

$$+ |x-a|_{\nu}^{rp_o(\alpha)p_o(\mu)p(\beta)p(\rho)} |y-b|_{\nu}^{rp_o(\alpha)p_o(\mu)p(\beta)p(\rho)}] \le$$

$$||a||_{\mu}^{rp_o(\alpha)p(\beta)p(\nu)p(\rho)} ||b||_{\mu}^{rp_o(\alpha)p(\beta)p(\nu)p(\rho)} + C^{rp_o(\mu)p(\nu)p(\rho)} \times$$

$$\times[C_1^{rp_o(\alpha)p(\beta)p(\rho)} ||a||_{\mu}^{rp_o(\alpha)p(\nu)p(\beta)p(\rho)} |y-b|_{\nu}^{rp_o(\alpha)p_o(\mu)p(\beta)p(\rho)} +$$

$$+C_1^{rp_o(\alpha)p(\beta)p(\rho)} |x-a|_{\nu}^{rp_o(\alpha)p_o(\mu)p(\beta)p(\rho)} ||b||_{\mu}^{rp_o(\alpha)p(\nu)p(\beta)p(\rho)} +$$

$$+|x-a|_{\nu}^{rp_o(\alpha)p_o(\mu)p(\beta)p(\rho)} |y-b|_{\nu}^{rp_o(\alpha)p_o(\mu)p(\beta)p(\rho)} \le$$

$$\le [||a||_{\mu}^{rp_o(\alpha)p(\beta)p(\nu)p(\rho)} + \tilde{C}^{rp_o(\alpha)p(\beta)p(\nu)} |x-a|_{\nu}^{rp_o(\alpha)p(\beta)p_o(\mu)p(\rho)}] \times$$

$$[||b||_{\mu}^{rp_o(\alpha)p(\beta)p(\nu)p(\rho)} + \tilde{C}^{rp_o(\alpha)p(\beta)p(\nu)} |y-b|_{\nu}^{rp_o(\alpha)p(\beta)p_o(\mu)p(\rho)}] \times$$

$$\le [||a||_{\mu}^{rp_o(\alpha)p(\beta)p(\nu)p(\rho)} + \tilde{C}^{rp_o(\alpha)p(\beta)p(\nu)} |x-a|_{\rho}^{rp_o(\alpha)p(\beta)p(\nu)p_o(\mu)}] \times$$

$$\times [||b||_{\mu}^{rp_o(\alpha)p(\beta)p(\nu)p(\rho)} + \tilde{C}^{rp_o(\alpha)p(\beta)p(\nu)} |y-b|_{\rho}^{rp_o(\alpha)p(\beta)p(\nu)p_o(\mu)}],$$

where $\tilde{C}$ is chosen so that $\tilde{C} \ge C_2$, $\tilde{C}^{2p_o(\alpha)p(\beta)} \ge C^{p_o(\mu)p(\rho)}$ and

$$\tilde{C}^{p_o(\alpha)p(\beta)p(\nu)} \ge C^{p_o(\mu)p(\nu)p(\rho)} \cdot C_1^{p_o(\alpha)p(\beta)p(\rho)} .$$

This choice of $\tilde{C}$ implies in particular that the relation (18) is still valid if we replace there C_2 by $\tilde{C}$, and so $\gamma_1 = (\mu,\rho, \tilde{C}, rp_o(\alpha)p(\beta)p(\nu))$ is in α_1. Taking in our inequality the infimum of the right-hand product with respect to a and b in A_o we obtain

$$(20) \qquad [|xy|_{\gamma}^{(1)}]^{p_o(\mu)p(\nu)p(\rho)} \le |x|_{\gamma_1}^{(1)} |y|_{\gamma_1}^{(1)}$$

Since $p_1(\gamma_1)=p_1(\gamma)p_o(\mu)p(\nu)p(\rho)$, then applying to both sides of (20) a power with exponent $p_1(\gamma)$ we obtain a relation of the form (5). Since γ was chosen arbitrarily in α_1 we conclude that the system ϕ_1 satisfies relations (5). The conclusion follows.

COROLLARY. The topology of a locally pseudoconvex algebra can be given by means of a family Φ of $p(\alpha)$-homogeneous seminorms satisfying relations (5) and (3).

REMARK. Our theorem cannot be replaced by the stronger one formulated exactly as the Fernández-Müller result. The reason is that a locally pseudoconvex algebra which is not locally convex may have a locally convex subalgebra, and the seminorms giving the topology of subalgebra cannot be extended to homogeneous seminorms giving the topology of the whole algebra.

REFERENCES

[1] A. Fernández , V.Müller, Renormalizations of Banach and locally convex algebras, Studia Math. /in print/

[2] S.Rolewicz, Metric linear spaces, PWN-Reidel, 1984.

[3] L.Waelbroeck, Topological vector spaces and algebras, Lecture Notes in Math. 230, Springer-Verlag 1971.

[4] W.Żelazko, Metric generalizations of Banach algebras, Rozprawy Mat. (Dissertationes Math.) 47 (1965).

[5] -, Selected topics in topological algebras, Aarhus Univ. Lecture Notes Series No. 31., 1971.

Mathematical Institute, Polish Academy of Sciences,
00-950 Warszawa (Poland), Sniadeckich 8.

Lecture Notes in Mathematics

For information about Vols. 1–1323
please contact your bookseller or Springer-Verlag

Vol. 1324: F. Cardoso, D.G. de Figueiredo, R. Iório, O. Lopes (Eds.), Partial Differential Equations. Proceedings, 1986. VIII, 433 pages. 1988.

Vol. 1325: A. Truman, I.M. Davies (Eds.), Stochastic Mechanics and Stochastic Processes. Proceedings, 1986. V, 220 pages. 1988.

Vol. 1326: P.S. Landweber (Ed.), Elliptic Curves and Modular Forms in Algebraic Topology. Proceedings, 1986. V, 224 pages. 1988.

Vol. 1327: W. Bruns, U. Vetter, Determinantal Rings. VII, 236 pages. 1988.

Vol. 1328: J.L. Bueso, P. Jara, B. Torrecillas (Eds.), Ring Theory. Proceedings, 1986. IX, 331 pages. 1988.

Vol. 1329: M. Alfaro, J.S. Dehesa, F.J. Marcellan, J.L. Rubio de Francia, J. Vinuesa (Eds.): Orthogonal Polynomials and their Applications. Proceedings, 1986. XV, 334 pages. 1988.

Vol. 1330: A. Ambrosetti, F. Gori, R. Lucchetti (Eds.), Mathematical Economics. Montecatini Terme 1986. Seminar. VII, 137 pages. 1988.

Vol. 1331: R. Bamón, R. Labarca, J. Palis Jr. (Eds.), Dynamical Systems, Valparaiso 1986. Proceedings. VI, 250 pages. 1988.

Vol. 1332: E. Odell, H. Rosenthal (Eds.), Functional Analysis. Proceedings. 1986–87. V, 202 pages. 1988.

Vol. 1333: A.S. Kechris, D.A. Martin, J.R. Steel (Eds.), Cabal Seminar 81–85. Proceedings, 1981–85. V, 224 pages. 1988.

Vol. 1334: Yu.G. Borisovich, Yu.E. Gliklikh (Eds.), Global Analysis – Studies and Applications III. V, 331 pages. 1988.

Vol. 1335: F. Guillén, V. Navarro Aznar, P. Pascual-Gainza, F. Puerta, Hyperrésolutions cubiques et descente cohomologique. XII, 192 pages. 1988.

Vol. 1336: B. Helffer, Semi-Classical Analysis for the Schrödinger Operator and Applications. V, 107 pages. 1988.

Vol. 1337: E. Sernesi (Ed.), Theory of Moduli. Seminar, 1985. VIII, 232 pages. 1988.

Vol. 1338: A.B. Mingarelli, S.G. Halvorsen. Non-Oscillation Domains of Differential Equations with Two Parameters. XI, 109 pages. 1988.

Vol. 1339: T. Sunada (Ed.), Geometry and Analysis of Manifolds. Proceedings, 1987. IX, 277 pages. 1988.

Vol. 1340: S. Hildebrandt, D.S. Kinderlehrer, M. Miranda (Eds.), Calculus of Variations and Partial Differential Equations. Proceedings, 1986. IX, 301 pages. 1988.

Vol. 1341: M. Dauge, Elliptic Boundary Value Problems on Corner Domains. VIII, 259 pages. 1988.

Vol. 1342: J.C. Alexander (Ed.), Dynamical Systems. Proceedings, 1986–87. VIII, 726 pages. 1988.

Vol. 1343: H. Ulrich, Fixed Point Theory of Parametrized Equivariant Maps. VII, 147 pages. 1988.

Vol. 1344: J. Král, J. Lukes, J. Netuka, J. Vesely´ (Eds.), Potential Theory – Surveys and Problems. Proceedings, 1987. VIII, 271 pages. 1988.

Vol. 1345: X. Gomez-Mont, J. Seade, A. Verjovski (Eds.), Holomorphic Dynamics. Proceedings, 1986. VII. 321 pages. 1988.

Vol. 1346: O.Ya. Viro (Ed.), Topology and Geometry – Rohlin Seminar. XI, 581 pages. 1988.

Vol. 1347: C. Preston, Iterates of Piecewise Monotone Mappings on an Interval. V, 166 pages. 1988.

Vol. 1348: F. Borceux (Ed.), Categorical Algebra and its Applications. Proceedings, 1987. VIII, 375 pages. 1988.

Vol. 1349: E. Novak, Deterministic and Stochastic Error Bounds in Numerical Analysis. V, 113 pages. 1988.

Vol. 1350: U. Koschorke (Ed.), Differential Topology Proceedings, 1987, VI, 269 pages. 1988.

Vol. 1351: I. Laine, S. Rickman, T. Sorvali (Eds.), Complex Analysis, Joensuu 1987. Proceedings. XV, 378 pages. 1988.

Vol. 1352: L.L. Avramov, K.B. Tchakerian (Eds.), Algebra – Some Current Trends. Proceedings. 1986. IX, 240 Seiten. 1988.

Vol. 1353: R.S. Palais, Ch.-l. Teng, Critical Point Theory and Submanifold Geometry. X, 272 pages. 1988.

Vol. 1354: A. Gómez, F. Guerra, M.A. Jiménez, G. López (Eds.), Approximation and Optimization. Proceedings, 1987. VI, 280 pages. 1988.

Vol. 1355: J. Bokowski, B. Sturmfels, Computational Synthetic Geometry. V, 168 pages. 1989.

Vol. 1356: H. Volkmer, Multiparameter Eigenvalue Problems and Expansion Theorems. VI, 157 pages. 1988.

Vol. 1357: S. Hildebrandt, R. Leis (Eds.), Partial Differential Equations and Calculus of Variations. VI, 423 pages. 1988.

Vol. 1358: D. Mumford, The Red Book of Varieties and Schemes. V, 309 pages. 1988.

Vol. 1359: P. Eymard, J.-P. Pier (Eds.) Harmonic Analysis. Proceedings, 1987. VIII, 287 pages. 1988.

Vol. 1360: G. Anderson, C. Greengard (Eds.), Vortex Methods. Proceedings, 1987. V, 141 pages. 1988.

Vol. 1361: T. tom Dieck (Ed.), Algebraic Topology and Transformation Groups. Proceedings. 1987. VI, 298 pages. 1988.

Vol. 1362: P. Diaconis, D. Elworthy, H. Föllmer, E. Nelson, G.C. Papanicolaou, S.R.S. Varadhan. École d´ Été de Probabilités de Saint-Flour XV–XVII. 1985–87 Editor: P.L. Hennequin. V, 459 pages. 1988.

Vol. 1363: P.G. Casazza, T.J. Shura, Tsirelson´s Space. VIII, 204 pages. 1988.

Vol. 1364: R.R. Phelps, Convex Functions, Monotone Operators and Differentiability. IX, 115 pages. 1989.

Vol. 1365: M. Giaquinta (Ed.), Topics in Calculus of Variations. Seminar, 1987. X, 196 pages. 1989.

Vol. 1366: N. Levitt, Grassmannians and Gauss Maps in PL-Topology. V, 203 pages. 1989.

Vol. 1367: M. Knebusch, Weakly Semialgebraic Spaces. XX, 376 pages. 1989.

Vol. 1368: R. Hübl, Traces of Differential Forms and Hochschild Homology. III, 111 pages. 1989.

Vol. 1369: B. Jiang, Ch.-K. Peng, Z. Hou (Eds.), Differential Geometry and Topology. Proceedings, 1986–87. VI, 366 pages. 1989.

Vol. 1370: G. Carlsson, R.L. Cohen, H.R. Miller, D.C. Ravenel (Eds.), Algebraic Topology. Proceedings, 1986. IX, 456 pages. 1989.

Vol. 1371: S. Glaz, Commutative Coherent Rings. XI, 347 pages. 1989.

Vol. 1372: J. Azéma, P.A. Meyer, M. Yor (Eds.), Séminaire de Probabilités XXIII. Proceedings. IV, 583 pages. 1989.

Vol. 1373: G. Benkart, J.M. Osborn (Eds.), Lie Algebras. Madison 1987. Proceedings. V, 145 pages. 1989.

Vol. 1374: R.C. Kirby, The Topology of 4-Manifolds. VI, 108 pages. 1989.

Vol. 1375: K. Kawakubo (Ed.), Transformation Groups. Proceedings, 1987. VIII, 394 pages, 1989.

Vol. 1376: J. Lindenstrauss, V.D. Milman (Eds.), Geometric Aspects of Functional Analysis. Seminar (GAFA) 1987–88. VII, 288 pages. 1989.

Vol. 1377: J.F. Pierce, Singularity Theory, Rod Theory, and Symmetry-Breaking Loads. IV, 177 pages. 1989.

Vol. 1378: R.S. Rumely, Capacity Theory on Algebraic Curves. III, 437 pages. 1989.

Vol. 1379: H. Heyer (Ed.), Probability Measures on Groups IX. Proceedings, 1988. VIII, 437 pages. 1989.

Vol. 1380: H.P. Schlickewei, E. Wirsing (Eds.), Number Theory, Ulm 1987. Proceedings. V, 266 pages. 1989.

Vol. 1381: J.-O. Strömberg, A. Torchinsky, Weighted Hardy Spaces. V, 193 pages. 1989.

Vol. 1382: H. Reiter, Metaplectic Groups and Segal Algebras. XI, 128 pages. 1989.

Vol. 1383: D.V. Chudnovsky, G.V. Chudnovsky, H. Cohn, M.B. Nathanson (Eds.), Number Theory, New York 1985–88. Seminar. V, 256 pages. 1989.

Vol. 1384: J. Garcia-Cuerva (Ed.), Harmonic Analysis and Partial Differential Equations. Proceedings, 1987. VII, 213 pages. 1989.

Vol. 1385: A.M. Anile, Y. Choquet-Bruhat (Eds.), Relativistic Fluid Dynamics. Seminar, 1987. V, 308 pages. 1989.

Vol. 1386: A. Bellen, C.W. Gear, E. Russo (Eds.), Numerical Methods for Ordinary Differential Equations. Proceedings, 1987. VII, 136 pages. 1989.

Vol. 1387: M. Petkovi´c, Iterative Methods for Simultaneous Inclusion of Polynomial Zeros. X, 263 pages. 1989.

Vol. 1388: J. Shinoda, T.A. Slaman, T. Tugué (Eds.), Mathematical Logic and Applications. Proceedings, 1987. V, 223 pages. 1989.

Vol. 1000: Second Edition. H. Hopf, Differential Geometry in the Large. VII, 184 pages. 1989.

Vol. 1389: E. Ballico, C. Ciliberto (Eds.), Algebraic Curves and Projective Geometry. Proceedings, 1988. V, 288 pages. 1989.

Vol. 1390: G. Da Prato, L. Tubaro (Eds.), Stochastic Partial Differential Equations and Applications II. Proceedings, 1988. VI, 258 pages. 1989.

Vol. 1391: S. Cambanis, A. Weron (Eds.), Probability Theory on Vector Spaces IV. Proceedings, 1987. VIII, 424 pages. 1989.

Vol. 1392: R. Silhol, Real Algebraic Surfaces. X, 215 pages. 1989.

Vol. 1393: N. Bouleau, D. Feyel, F. Hirsch, G. Mokobodzki (Eds.), Séminaire de Théorie du Potentiel Paris, No. 9. Proceedings. VI, 265 pages. 1989.

Vol. 1394: T.L. Gill, W.W. Zachary (Eds.), Nonlinear Semigroups, Partial Differential Equations and Attractors. Proceedings, 1987. IX, 233 pages. 1989.

Vol. 1395: K. Alladi (Ed.), Number Theory, Madras 1987. Proceedings. VII, 234 pages. 1989.

Vol. 1396: L. Accardi, W. von Waldenfels (Eds.), Quantum Probability and Applications IV. Proceedings, 1987. VI, 355 pages. 1989.

Vol. 1397: P.R. Turner (Ed.), Numerical Analysis and Parallel Processing. Seminar, 1987. VI, 264 pages. 1989.

Vol. 1398: A.C. Kim, B.H. Neumann (Eds.), Groups – Korea 1988. Proceedings. V, 189 pages. 1989.

Vol. 1399: W.-P. Barth, H. Lange (Eds.), Arithmetic of Complex Manifolds. Proceedings, 1988. V, 171 pages. 1989.

Vol. 1400: U. Jannsen. Mixed Motives and Algebraic K-Theory. XIII, 246 pages. 1990.

Vol. 1401: J. Steprans, S. Watson (Eds.), Set Theory and its Applications. Proceedings, 1987. V, 227 pages. 1989.

Vol. 1402: C. Carasso, P. Charrier, B. Hanouzet, J.-L. Joly (Eds.), Nonlinear Hyperbolic Problems. Proceedings, 1988. V, 249 pages. 1989.

Vol. 1403: B. Simeone (Ed.), Combinatorial Optimization. Seminar, 1986. V, 314 pages. 1989.

Vol. 1404: M.-P. Malliavin (Ed.), Séminaire d´AIgèbre Paul Dubreil et Marie-Paul Malliavin. Proceedings, 1987–1988. IV, 410 pages. 1989.

Vol. 1405: S. Dolecki (Ed.), Optimization. Proceedings, 1988. V, 223 pages. 1989. Vol. 1406: L. Jacobsen (Ed.), Analytic Theory of Continued Fractions III. Proceedings, 1988. VI, 142 pages. 1989.

Vol. 1407: W. Pohlers, Proof Theory. VI, 213 pages. 1989.

Vol. 1408: W. Lück, Transformation Groups and Algebraic K-Theory. XII, 443 pages. 1989.

Vol. 1409: E. Hairer, Ch. Lubich, M. Roche. The Numerical Solution of Differential-Algebraic Systems by Runge-Kutta Methods. VII, 139 pages. 1989.

Vol. 1410: F.J. Carreras, O. Gil-Medrano, A.M. Naveira (Eds.), Differential Geometry. Proceedings, 1988. V, 308 pages. 1989.

Vol. 1411: B. Jiang (Ed.), Topological Fixed Point Theory and Applications. Proceedings. 1988. VI, 203 pages. 1989.

Vol. 1412: V.V. Kalashnikov, V.M. Zolotarev (Eds.), Stability Problems for Stochastic Models. Proceedings, 1987. X, 380 pages. 1989.

Vol. 1413: S. Wright, Uniqueness of the Injective lll_1Factor. III, 108 pages. 1989.

Vol. 1414: E. Ramirez de Arellano (Ed.), Algebraic Geometry and Complex Analysis. Proceedings, 1987. VI, 180 pages. 1989.

Vol. 1415: M. Langevin, M. Waldschmidt (Eds.), Cinquante Ans de Polynômes. Fifty Years of Polynomials. Proceedings, 1988. IX, 235 pages.1990.

Vol. 1416: C. Albert (Ed.), Géométrie Symplectique et Mécanique. Proceedings, 1988. V, 289 pages. 1990.

Vol. 1417: A.J. Sommese, A. Biancofiore, E.L. Livorni (Eds.), Algebraic Geometry. Proceedings, 1988. V, 320 pages. 1990.

Vol. 1418: M. Mimura (Ed.), Homotopy Theory and Related Topics. Proceedings, 1988. V, 241 pages. 1990.

Vol. 1419: P.S. Bullen, P.Y. Lee, J.L. Mawhin, P. Muldowney, W.F. Pfeffer (Eds.), New Integrals. Proceedings, 1988. V, 202 pages. 1990.

Vol. 1420: M. Galbiati, A. Tognoli (Eds.), Real Analytic Geometry. Proceedings, 1988. IV, 366 pages. 1990.

Vol. 1421: H.A. Biagioni, A Nonlinear Theory of Generalized Functions, XII, 214 pages. 1990.

Vol. 1422: V. Villani (Ed.), Complex Geometry and Analysis. Proceedings, 1988. V, 109 pages. 1990.

Vol. 1423: S.O. Kochman, Stable Homotopy Groups of Spheres: A Computer-Assisted Approach. VIII, 330 pages. 1990.

Vol. 1424: F.E. Burstall, J.H. Rawnsley, Twistor Theory for Riemannian Symmetric Spaces. III, 112 pages. 1990.

Vol. 1425: R.A. Piccinini (Ed.), Groups of Self-Equivalences and Related Topics. Proceedings, 1988. V, 214 pages. 1990.

Vol. 1426: J. Azéma, P.A. Meyer, M. Yor (Eds.), Séminaire de Probabilités XXIV, 1988/89. V, 490 pages. 1990.

Vol. 1427: A. Ancona, D. Geman, N. Ikeda, École d'Eté de Probabilités de Saint Flour XVIII, 1988. Ed.: P.L. Hennequin. VII, 330 pages. 1990.

Vol. 1428: K. Erdmann, Blocks of Tame Representation Type and Related Algebras. XV. 312 pages. 1990.

Vol. 1429: S. Homer, A. Nerode, R.A. Platek, G.E. Sacks, A. Scedrov, Logic and Computer Science. Seminar, 1988. Editor: P. Odifreddi. V, 162 pages. 1990.

Vol. 1430: W. Bruns, A. Simis (Eds.), Commutative Algebra. Proceedings. 1988. V, 160 pages. 1990.

Vol. 1431: J.G. Heywood, K. Masuda, R. Rautmann, V.A. Solonnikov (Eds.), The Navier-Stokes Equations – Theory and Numerical Methods. Proceedings, 1988. VII, 238 pages. 1990.

Vol. 1432: K. Ambos-Spies, G.H. Müller, G.E. Sacks (Eds.), Recursion Theory Week. Proceedings, 1989. VI, 393 pages. 1990.

Vol. 1433: S. Lang, W. Cherry, Topics in Nevanlinna Theory. II, 174 pages.1990.

Vol. 1434: K. Nagasaka, E. Fouvry (Eds.), Analytic Number Theory. Proceedings, 1988. VI, 218 pages. 1990.

Vol. 1435: St. Ruscheweyh, E.B. Saff, L.C. Salinas, R.S. Varga (Eds.), Computational Methods and Function Theory. Proceedings, 1989. VI, 211 pages. 1990.

Vol. 1436: S. Xambó-Descamps (Ed.), Enumerative Geometry. Proceedings, 1987. V, 303 pages. 1990.

Vol. 1437: H. Inassaridze (Ed.), K-theory and Homological Algebra. Seminar, 1987–88. V, 313 pages. 1990.

Vol. 1438: P.G. Lemarié (Ed.) Les Ondelettes en 1989. Seminar. IV, 212 pages. 1990.

Vol. 1439: E. Bujalance, J.J. Etayo, J.M. Gamboa, G. Gromadzki. Automorphism Groups of Compact Bordered Klein Surfaces: A Combinatorial Approach. XIII, 201 pages. 1990.

Vol. 1440: P. Latiolais (Ed.), Topology and Combinatorial Groups Theory. Seminar, 1985–1988. VI, 207 pages. 1990.

Vol. 1441: M. Coornaert, T. Delzant, A. Papadopoulos. Géométrie et théorie des groupes. X, 165 pages. 1990.

Vol. 1442: L. Accardi, M. von Waldenfels (Eds.), Quantum Probability and Applications V. Proceedings, 1988. VI, 413 pages. 1990.

Vol. 1443: K.H. Dovermann, R. Schultz, Equivariant Surgery Theories and Their Periodicity Properties. VI, 227 pages. 1990.

Vol. 1444: H. Korezlioglu, A.S. Ustunel (Eds.), Stochastic Analysis and Related Topics VI. Proceedings, 1988. V, 268 pages. 1990.

Vol. 1445: F. Schulz, Regularity Theory for Quasilinear Elliptic Systems and – Monge Ampère Equations in Two Dimensions. XV, 123 pages. 1990.

Vol. 1446: Methods of Nonconvex Analysis. Seminar, 1989. Editor: A. Cellina. V, 206 pages. 1990.

Vol. 1447: J.-G. Labesse, J. Schwermer (Eds), Cohomology of Arithmetic Groups and Automorphic Forms. Proceedings, 1989. V, 358 pages. 1990.

Vol. 1448: S.K. Jain, S.R. López-Permouth (Eds.), Non-Commutative Ring Theory. Proceedings, 1989. V, 166 pages. 1990.

Vol. 1449: W. Odyniec, G. Lewicki, Minimal Projections in Banach Spaces. VIII, 168 pages. 1990.

Vol. 1450: H. Fujita, T. Ikebe, S.T. Kuroda (Eds.), Functional-Analytic Methods for Partial Differential Equations. Proceedings, 1989. VII, 252 pages. 1990.

Vol. 1451: L. Alvarez-Gaumé, E. Arbarello, C. De Concini, N.J. Hitchin, Global Geometry and Mathematical Physics. Montecatini Terme 1988. Seminar. Editors: M. Francaviglia, F. Gherardelli. IX, 197 pages. 1990.

Vol. 1452: E. Hlawka, R.F. Tichy (Eds.), Number-Theoretic Analysis. Seminar, 1988–89. V, 220 pages. 1990.

Vol. 1453: Yu.G. Borisovich, Yu.E. Gliklikh (Eds.), Global Analysis – Studies and Applications IV. V, 320 pages. 1990.

Vol. 1454: F. Baldassari, S. Bosch, B. Dwork (Eds.), p-adic Analysis. Proceedings, 1989. V, 382 pages. 1990.

Vol. 1455: J.-P. Françoise, R. Roussarie (Eds.), Bifurcations of Planar Vector Fields. Proceedings, 1989. VI, 396 pages. 1990.

Vol. 1456: L.G. Kovács (Ed.), Groups – Canberra 1989. Proceedings. XII, 198 pages. 1990.

Vol. 1457: O. Axelsson, L.Yu. Kolotilina (Eds.), Preconditioned Conjugate Gradient Methods. Proceedings, 1989. V, 196 pages. 1990.

Vol. 1458: R. Schaaf, Global Solution Branches of Two Point Boundary Value Problems. XIX, 141 pages. 1990.

Vol. 1459: D. Tiba, Optimal Control of Nonsmooth Distributed Parameter Systems. VII, 159 pages. 1990.

Vol. 1460: G. Toscani, V. Boffi, S. Rionero (Eds.), Mathematical Aspects of Fluid Plasma Dynamics. Proceedings, 1988. V, 221 pages. 1991.

Vol. 1461: R. Gorenflo, S. Vessella, Abel Integral Equations. VII, 215 pages. 1991.

Vol. 1462: D. Mond, J. Montaldi (Eds.), Singularity Theory and its Applications. Warwick 1989, Part I. VIII, 405 pages. 1991.

Vol. 1463: R. Roberts, I. Stewart (Eds.), Singularity Theory and its Applications. Warwick 1989, Part II. VIII, 322 pages. 1991.

Vol. 1464: D. L. Burkholder, E. Pardoux, A. Sznitman, Ecole d'Eté de Probabilités de Saint- Flour XIX-1989. Editor: P. L. Hennequin. VI, 256 pages. 1991.

Vol. 1465: G. David, Wavelets and Singular Integrals on Curves and Surfaces. X, 107 pages. 1991.

Vol. 1466: W. Banaszczyk, Additive Subgroups of Topological Vector Spaces. VII, 178 pages. 1991.

Vol. 1467: W. M. Schmidt, Diophantine Approximations and Diophantine Equations. VIII, 217 pages. 1991.

Vol. 1468: J. Noguchi, T. Ohsawa (Eds.), Prospects in Complex Geometry. Proceedings, 1989. VII, 421 pages. 1991.

Vol. 1469: J. Lindenstrauss, V. D. Milman (Eds.), Geometric Aspects of Functional Analysis. Seminar 1989-90. XI, 191 pages. 1991.

Vol. 1470: E. Odell, H. Rosenthal (Eds.), Functional Analysis. Proceedings, 1987-89. VII, 199 pages. 1991.

Vol. 1471: A. A. Panchishkin, Non-Archimedean L-Functions of Siegel and Hilbert Modular Forms. VII, 157 pages. 1991.

Vol. 1472: T. T. Nielsen, Bose Algebras: The Complex and Real Wave Representations. V, 132 pages. 1991.

Vol. 1473: Y. Hino, S. Murakami, T. Naito, Functional Differential Equations with Infinite Delay. X, 317 pages. 1991.

Vol. 1474: S. Jackowski, B. Oliver, K. Pawałowski (Eds.), Algebraic Topology, Poznań 1989. Proceedings. VIII, 397 pages. 1991.

Vol. 1475: S. Busenberg, M. Martelli (Eds.), Delay Differential Equations and Dynamical Systems. Proceedings, 1990. VIII, 249 pages. 1991.

Vol. 1476: M. Bekkali, Topics in Set Theory. VII, 120 pages. 1991.

Vol. 1477: R. Jajte, Strong Limit Theorems in Noncommutative L_2-Spaces. X, 113 pages. 1991.

Vol. 1478: M.-P. Malliavin (Ed.), Topics in Invariant Theory. Seminar 1989-1990. VI, 272 pages. 1991.

Vol. 1479: S. Bloch, I. Dolgachev, W. Fulton (Eds.), Algebraic Geometry. Proceedings, 1989. VII, 300 pages. 1991.

Vol. 1480: F. Dumortier, R. Roussarie, J. Sotomayor, H. Żołądek, Bifurcations of Planar Vector Fields: Nilpotent Singularities and Abelian Integrals. VIII, 226 pages. 1991.

Vol. 1481: D. Ferus, U. Pinkall, U. Simon, B. Wegner (Eds.), Global Differential Geometry and Global Analysis. Proceedings, 1991. VIII, 283 pages. 1991.

Vol. 1482: J. Chabrowski, The Dirichlet Problem with L^2-Boundary Data for Elliptic Linear Equations. VI, 173 pages. 1991.

Vol. 1483: E. Reithmeier, Periodic Solutions of Nonlinear Dynamical Systems. VI, 171 pages. 1991.

Vol. 1484: H. Delfs, Homology of Locally Semialgebraic Spaces. IX, 136 pages. 1991.

Vol. 1485: J. Azéma, P. A. Meyer, M. Yor (Eds.), Séminaire de Probabilités XXV. VIII, 440 pages. 1991.

Vol. 1486: L. Arnold, H. Crauel, J.-P. Eckmann (Eds.), Lyapunov Exponents. Proceedings, 1990. VIII, 365 pages. 1991.

Vol. 1487: E. Freitag, Singular Modular Forms and Theta Relations. VI, 172 pages. 1991.

Vol. 1488: A. Carboni, M. C. Pedicchio, G. Rosolini (Eds.), Category Theory. Proceedings, 1990. VII, 494 pages. 1991.

Vol. 1489: A. Mielke, Hamiltonian and Lagrangian Flows on Center Manifolds. X, 140 pages. 1991.

Vol. 1490: K. Metsch, Linear Spaces with Few Lines. XIII, 196 pages. 1991.

Vol. 1491: E. Lluis-Puebla, J.-L. Loday, H. Gillet, C. Soulé, V. Snaith, Higher Algebraic K-Theory: an overview. IX, 164 pages. 1992.

Vol. 1492: K. R. Wicks, Fractals and Hyperspaces. VIII, 168 pages. 1991.

Vol. 1493: E. Benoît (Ed.), Dynamic Bifurcations. Proceedings, Luminy 1990. VII, 219 pages. 1991.

Vol. 1494: M.-T. Cheng, X.-W. Zhou, D.-G. Deng (Eds.), Harmonic Analysis. Proceedings, 1988. IX, 226 pages. 1991.

Vol. 1495: J. M. Bony, G. Grubb, L. Hörmander, H. Komatsu, J. Sjöstrand, Microlocal Analysis and Applications. Montecatini Terme, 1989. Editors: L. Cattabriga, L. Rodino. VII, 349 pages. 1991.

Vol. 1496: C. Foias, B. Francis, J. W. Helton, H. Kwakernaak, J. B. Pearson, H_∞-Control Theory. Como, 1990. Editors: E. Mosca, L. Pandolfi. VII, 336 pages. 1991.

Vol. 1497: G. T. Herman, A. K. Louis, F. Natterer (Eds.), Mathematical Methods in Tomography. Proceedings 1990. X, 268 pages. 1991.

Vol. 1498: R. Lang, Spectral Theory of Random Schröding Operators. X, 125 pages. 1991.

Vol. 1499: K. Taira, Boundary Value Problems and Mark Processes. IX, 132 pages. 1991.

Vol. 1500: J.-P. Serre, Lie Algebras and Lie Groups. VII, 1 pages. 1992.

Vol. 1501: A. De Masi, E. Presutti, Mathematical Methods f Hydrodynamic Limits. IX, 196 pages. 1991.

Vol. 1502: C. Simpson, Asymptotic Behavior of Monodrom V, 139 pages. 1991.

Vol. 1503: S. Shokranian, The Selberg-Arthur Trace Formu (Lectures by J. Arthur). VII, 97 pages. 1991.

Vol. 1504: J. Cheeger, M. Gromov, C. Okonek, P. Pans Geometric Topology: Recent Developments. Editors: P. Bartolomeis, F. Tricerri. VII, 197 pages. 1991.

Vol. 1505: K. Kajitani, T. Nishitani, The Hyperbolic Caucl Problem. VII, 168 pages. 1991.

Vol. 1506: A. Buium, Differential Algebraic Groups of Fini Dimension. XV, 145 pages. 1992.

Vol. 1507: K. Hulek, T. Peternell, M. Schneider, F.-O. Schrey (Eds.), Complex Algebraic Varieties. Proceedings, 1990. V 179 pages. 1992.

Vol. 1508: M. Vuorinen (Ed.), Quasiconformal Space Mapping A Collection of Surveys 1960-1990. IX, 148 pages. 1992.

Vol. 1509: J. Aguadé, M. Castellet, F. R. Cohen (Eds.), Algebra Topology - Homotopy and Group Cohomology. Proceeding 1990. X, 330 pages. 1992.

Vol. 1510: P. P. Kulish (Ed.), Quantum Groups. Proceeding 1990. XII, 398 pages. 1992.

Vol. 1511: B. S. Yadav, D. Singh (Eds.), Functional Analys and Operator Theory. Proceedings, 1990. VIII, 223 pages. 199